Greenhouse Earth
Tomorrow's Disaster Today

Jon Erickson

TAB BOOKS
Blue Ridge Summit, PA

FIRST EDITION
SECOND PRINTING

© 1990 by **TAB Books**.
TAB Books is a division of McGraw-Hill, Inc.

Library of Congress Cataloging-in-Publication Data

Erickson, Jon, 1948-
    Greenhouse earth : tomorrow's disaster today / by Jon Erickson
      p.   cm.
    Includes bibliographical references.
    ISBN 0-8306-8471-9   ISBN 0-8306-3471-1 (pbk.)
    1. Global warming. 2. Greenhouse effect, Atmospheric. 3. Man-
-Influence on nature. I. Title.
QC981.8.G56E75   1990
363.73'87—dc20   89-77650
                                                          CIP

TAB Books offers software for sale. For information and a catalog, please contact
TAB Software Department, Blue Ridge Summit, PA 17294-0850.

Acquisitions Editor: Roland S. Phelps
Technical Editor: Barbara B. Minich
Production: Katherine G. Brown
Cover photograph courtesy of NASA.   HT3

Contents

Acknowledgments

I wish to thank the following organizations for providing photographs for this book: the Department of Agriculture Soil Conservation Service, the National Aeronautics and Space Administration, the National Oceanic and Atmospheric Administration, the National Park Service, the U.S. Army Corps of Engineers, the U.S. Coast Guard, the U.S. Department of Energy, the U.S. Navy, and the U.S. Geological Survey.

Introduction

For the last 10,000 years, the Earth's climate has been extraordinarily beneficial to mankind. Humans have prospered exceedingly well under a benign atmosphere. Today, however, major changes are taking place. Human beings are conducting an inadvertent global experiment by changing the face of the entire planet. We are destroying the rain forests and pumping our pollutants into the air and water. Some of these pollutants are extremely toxic and carcinogenic. Others are destroying the ozone layer, which allows life to exist on the Earth's surface. All these activities are unfavorably altering the composition of the biosphere and the Earth's heat balance.

If we do not curb our insatiable appetite for fossil fuels and stop destroying the forests, the world could become hotter than it has been in the past million years. Average global temperatures have risen 1 degree Fahrenheit over the last century. If carbon dioxide and other greenhouse gasses continue to spill into the atmosphere, global temperatures could rise 5 to 10 degrees by the middle of the next century. The warning will be greatest at the higher latitudes of the Northern Hemisphere, with the largest temperature increases occurring in winter. Most areas will experience summertime highs well above 100 degrees Fahrenheit. New temperature records will be set each year. As a possible prelude to global warming, the decade of the 1980s has had the six hottest years of the century.

Atmospheric disturbances brought on by the additional warming will produce more violent storms and larger death tolls. Some areas, particularly in the Northern Hemisphere, will dry out and a greater occurrence of lightning strikes will set massive forest fires. The charring of the Earth by natural and man-made forest fires will dump additional quantities of carbon dioxide into the atmosphere. Changes in temperature and rainfall brought on by global warming will in turn change the composition of the forests. At the present rate of destruction, most of the rain forests will be gone by the middle of the next century. This will allow man-made deserts to encroach on once lush areas.

Evaporation rates will also increase and circulation patterns will change. Decreased rainfall in some areas will result in increased rainfall in others. In some regions, river flow will be reduced or stopped completely. Other areas will experience sudden downpours that create massive floods. The central portions of the continents, which normally experience occasional droughts, might become permanently dry wastelands. Vast areas of once productive cropland could lose topsoil and become man-made deserts.

Coastal regions, where half the human population lives, will feel the adverse effects of rising sea levels as the ice caps melt under rising ocean temperatures. If the present melting continues, the sea could rise as much as 6 feet by the middle of the next century. Large tracks of coastal land would disappear, as would shallow barrier islands and coral reefs. Low-lying fertile deltas that support millions of people would vanish. Delicate wetlands, where many species of marine life hatch their young, would be reclaimed by the sea. Vulnerable coastal cities would have to move farther inland or build protective walls against an angry sea, where a larger number of extremely dangerous hurricanes would prowl the ocean stretches.

Forests and other wildlife habitats might not have enough time to adjust to the rapidly changing climate. The warming will rearrange entire biological communities and cause many species to become extinct. Weeds and pests could overrun much of the landscape.

Since life controls the climate to some extent, it is uncertain what long-term effects a diminished biosphere will have on the world as a whole. It is becoming more apparent, however, that as man continues to squander the Earth's resources, the climate could change in such a way that it is no longer benevolent to mankind.

1

The Primordial Greenhouse

IT IS not just a fortunate coincidence that the temperatures on Earth are between the extremes of Venus, which roasts at the temperature of molten lead, and Mars, which is as cold as Antarctica in the dead of winter. If Mars had Venus's heavy carbon dioxide atmosphere, it could actually be hotter than Earth because the greenhouse effect would retain the little amount of heat Mars receives from the Sun. On the other hand, if Venus had Mars's thin carbon dioxide atmosphere, it could actually be colder than Earth, even though it is closer to the Sun.

All life on Earth owes its existence to the greenhouse effect. Large quantities of greenhouse gasses in the Earth's early atmosphere kept the temperature within tolerable limits for life to flourish. During its infancy, the Sun's output was about a third less than it is today. The Earth then received about the same amount of sunlight at the equator that Antarctica now receives in summer. There is no geological evidence, however, that the Earth was entirely frozen over during this time. Actually, it was even warmer then than it is today. Carbon dioxide, which then comprised about one-quarter of the gasses in the atmosphere, kept the Earth balmy and allowed the initiation and growth of life.

For the last 2 billion years, the Earth has received a fairly steady amount of solar energy. Major changes in the climate have been brought on by changes in the carbon dioxide content of the atmosphere. When too much carbon dioxide was removed by the carbon cycle, temperatures plummeted and great ice sheets flowed across the land. When too much carbon dioxide was allowed to build up in the atmosphere due to excessive volcanic activity, the Earth became a hothouse. Only when carbon dioxide levels have held fairly constant has the climate been optimal for all living things.

THE FEEBLE SUN

About 4.6 billion years ago, the Solar System condensed out of an enormous cloud of gas and dust (FIG. 1-1). The Sun was extremely unstable during its first billion years. Its output was only about 70 percent of what it is today. The Sun's rays provided as much warmth to the Earth then as they now do on Mars. Periodically, intense nuclear reactions in the Sun's core created large thermal pressures, which caused the Sun to expand up to a third larger than its present size. The intense internal turmoil produced powerful solar flares that leaped millions of miles into space. This high-temperature plasma of atomic particles produced a hurricane-strength solar wind that was much stronger than the solar wind experienced today.

(COURTESY OF NASA)

Fig. 1-1. The Solar System condensed from a dense cloud similar to the Orion Nebula.

For a short time, all this activity made the Sun radiate more heat. This in turn cooled the core, and the Sun returned to its original size. The early Sun spun around on its axis much faster. A single rotation then took just a few days compared to the 27 days it now takes. This produced strong magnetic fields and considerable turmoil on the Sun's surface. The result was numerous giant sunspots and solar flares (FIG. 1-2). This produced energetic solar particles that shot out from the Sun's equator and bombarded the Earth with intense particle radiation.

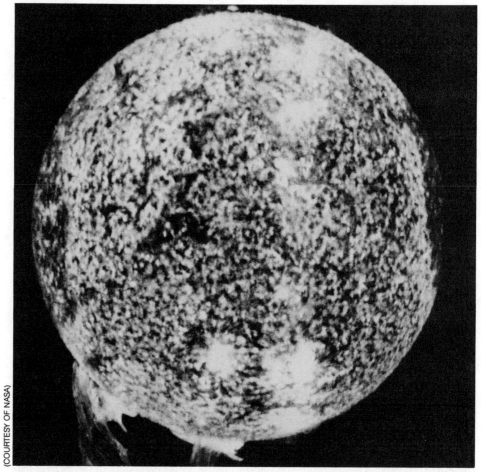

Fig. 1-2. Photograph of the Sun taken by Skylab *that shows one of the most spectacular solar flares ever recorded.*

By the time the Earth cooled enough to allow an ocean to form, the atmosphere had evolved into a mixture of water vapor, carbon dioxide, methane, unstable ammonia, which broke down into hydrogen and nitrogen, and traces of other gasses. The original atmosphere probably contained as much as 25 percent carbon dioxide. This gas is transparent to incoming sunlight, but absorbs and reradiates groundward outgoing infrared radiation, which creates a greenhouse effect. As a result, the early Earth was able to retain a good portion of the Sun's energy (FIG. 1-3).

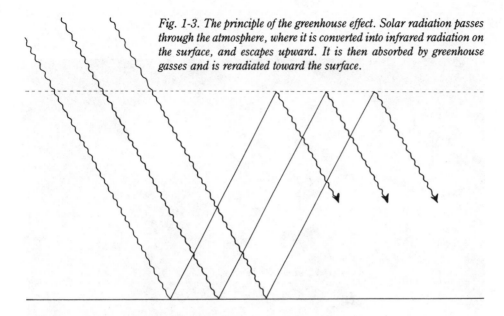

Fig. 1-3. The principle of the greenhouse effect. Solar radiation passes through the atmosphere, where it is converted into infrared radiation on the surface, and escapes upward. It is then absorbed by greenhouse gasses and is reradiated toward the surface.

If the early Earth had had the present atmosphere, the average global temperature would have been well below the freezing point of water and the entire ocean would have been a solid block of ice. If the Earth had entirely frozen over, it is doubtful that it could have thawed out, even if the Sun's energy had been 50 percent more than it is today. Because ice is a good reflector of sunlight, the Earth probably would have become a glistening orb similar to the icy moons of Jupiter and Saturn (FIG. 1-4).

On the other hand, if the Sun's output had been equal then to what it is today, the ocean would have boiled away. The large quantity of carbon dioxide in the atmosphere would have retained the Sun's heat and dramatically increased temperatures. The presence of abundant deuterium in seawater does suggest that at some time large quantities of water did escape the Earth.

As the Sun became progressively hotter, carbon dioxide was gradually taken out of the Earth's atmosphere and stored on the continents and ocean floors in the form of carbonate rocks. If the level of carbon dioxide had been allowed to increase without these moderating factors, the Earth would have become much too hot and life would have had little chance to evolve.

As time progressed, green plants dominated the planet. They removed carbon dioxide, combined it with sunlight, and stored it in their tissues as carbohydrates. The Earth probably cooled as a result of the evolution of surface plants because plants greatly accelerated the removal of carbon dioxide from the atmosphere and increased the concentration of carbon in the soil. Today, the amount of soil carbon is about 40 times greater than the level of carbon in the atmosphere.

The first forests grew in the great swamps that existed during the Carboniferous period, 350 to 280 million years ago (FIG. 1-5). Plant growth also was highly prodigious, probably due to abundant amounts of carbon dioxide in the atmosphere, which acted as a sort of fertilizer. Plants formed thick layers of vegetative matter that was later buried. Today's thick coal beds (FIG. 1-6) are a result of this buried matter.

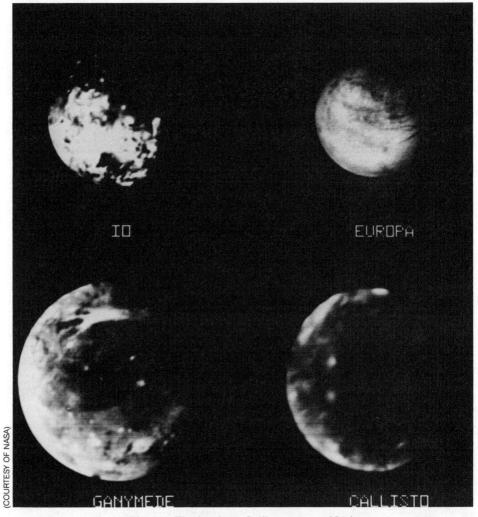

Fig. 1-4. The four large Galilean moons of Jupiter.

THE ORIGINAL ATMOSPHERE

The Solar System is comprised mostly of hydrogen and helium, which reside in the Sun and the atmospheres of the giant gaseous planets Jupiter, Saturn, Uranus, and Neptune. The gasses and dust particles that pervaded the early nebula were supplied by giant exploding stars called supernovas. During its formation, the Earth acquired a massive atmosphere comprised of hydrogen and helium, which it pulled from the solar nebula.

As the Earth orbited the Sun, it grew in size by acquiring small, stony bodies called planetesimals. As it grew, it slowed down and began to spiral inward toward the Sun. This placed the Earth in a lower orbit and allowed it to collide with more planetesimals until eventually it collected all the material in its path. The Earth's orbit then stabilized near where it is today. Gravitational compression of the atmosphere caused high surface temperatures. Surface temperatures were hot enough to melt rocks, much like the atmospheric conditions found on Venus today.

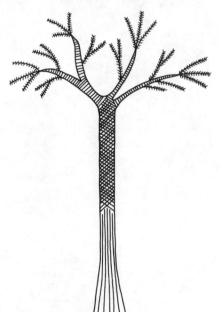

Fig. 1-5. The scale tree was one of many early trees in the lush Carboniferous forests.

Fig. 1-6. The Big Elk coal bed in King County, Washington.

(PHOTO BY J.D. VINE, COURTESY OF USGS)

The constant tug of the strong solar wind completely stripped away the Earth's atmosphere. Without an atmosphere, the planet lost its surface heat. Rocks cooled and a thin temporary crust formed. For the next several hundred million years, the Earth did not have an atmosphere. The surface was in a near vacuum state, much like the Moon is now. Without an atmosphere or ocean to distribute the Sun's heat, the ground baked during the day and froze at night. At that time, both day and night were only a few hours long and the Moon loomed so close that it filled much of the sky. The nearness of the Moon produced high tides on the Earth's flexible crust.

Because there was no atmosphere, volcanic debris shot straight up to great heights. Since it was not scattered by air currents, the debris fell back to Earth and landed around the volcanic vent. This created super volcanoes that were substantially taller than volcanoes are today. Fountains of lava burst through cracks in the Earth's thin crust and its surface was paved by floods of basalt, which formed maria, or basaltic plains. The entire surface of the Earth was dotted by mighty volcanoes that erupted in succession, and produced great blasts of fire.

The absence of an atmosphere also left the Earth vulnerable to meteorites. Between 4.2 and 3.9 billion years ago, the Earth was heavily bombarded by meteorites. The Moon, Mercury, Mars, and the moons of the large gaseous planets were also bombarded and still show signs of this massive attack (FIG. 1-7).

Soon after the start of the massive meteorite bombardment, the Earth began to slowly acquire a new atmosphere, which contained nitrogen, carbon dioxide, ammonia, methane, and water vapor. It became so saturated with water vapor that the atmospheric pressure was several times greater than it is today.

Most of the water vapor and gasses in the new atmosphere came from within the Earth itself. Magma contains large amounts of volatiles, mostly water and carbon dioxide, which help to make it fluid. The volatiles remain in the magma for as long as it stays deep inside the Earth. When the magma rises to the surface, however, the drop in pressure explosively releases the trapped water and gasses. The early volcanoes erupted much more violently than the present ones. This was because the Earth's interior was much hotter and the magma contained more volatiles than it does today.

Comets, which are essentially rocky material encased in ice, also plunged to the Earth and released large amounts of water vapor and gas (FIG. 1-8). These cosmic gasses were mostly carbon dioxide, ammonia, and methane.

Oxygen was produced directly by volcanic outgassing and the degassing of meteorites and indirectly by the breakdown of water vapor and carbon dioxide molecules from ultraviolet radiation. All oxygen generated in this manner was quickly bonded to metals in the crust, much like oxygen reacts with iron to make rust. Oxygen also combined with hydrogen and carbon monoxide to reconstitute water vapor and carbon dioxide. There might also have been a small amount of oxygen in the upper atmosphere, where it could have provided a thin ozone screen. This might have reduced the breakdown of water molecules from ultraviolet rays and prevented the Earth's ocean from evaporating entirely.

Nitrogen, which makes up about 80 percent of the atmosphere, is practically inert and does not contribute to the greenhouse effect. It originally came from volcanic eruptions (FIG. 1-9) and the breakdown of ammonia, which consists of one nitrogen atom and three hydrogen atoms. Unlike most other gasses, which have been replaced or recycled, the Earth still contains much of its original nitrogen. This is because life prevents all the nitrogen from being transformed into nitrate, which is easily dissolved in the ocean. Bacteria in the ocean return the nitrate-nitrogen to its gaseous state. If this were not so, the Earth would have only a fraction of its present atmospheric pressure.

Fig. 1-7. View of a full Moon showing numerous craters taken from Apollo 11 *spacecraft.*

(COURTESY OF NASA)

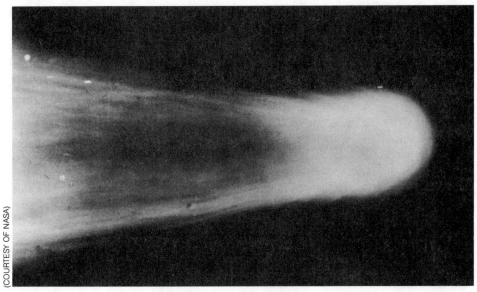

Fig. 1-8. Halley's Comet.

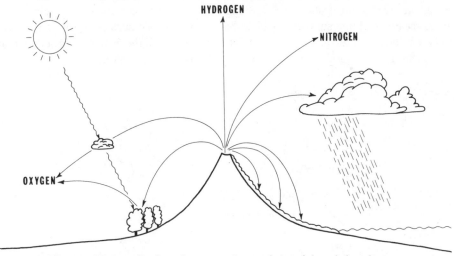

Fig. 1-9. The contribution of gasses and crustal material made by volcanoes.

ORIGIN OF THE OCEAN

As the atmosphere was forming, the entire surface of the Earth was in a constant rage. Winds blew with such force that present-day hurricanes would have seemed like gentle breezes. Dust storms were prevalent on the dry surface. They covered the entire planet with suspended sediment much like the Martian dust storms. Huge lightning bolts flashed constantly across the sky. Thunder caused one gigantic shock wave after another to reverberate across the land.

The restless Earth was rent apart by massive quakes that cracked open the thin crust. Magma flowed through the fissures and paved over the surface with lava, forming flat featureless plains. Volcanoes erupted in one giant outburst after another. The intense volcanic activity lofted millions of tons of volcanic debris into the atmosphere, where it remained suspended for long periods. The volcanic ash and dust particles scattered the sunlight and gave an eerie red glow to the sky.

The dust also cooled the Earth and acted as nuclei upon which water vapor could condense. When the upper atmospheric temperature lowered to the dew point, water vapor condensed into cloud droplets. Clouds became so thick and heavy they almost completely blocked out the Sun.

The lack of sunlight allowed surface temperatures to drop considerably. Further cooling of the atmosphere allowed rain drops to form. The Earth received deluge after deluge as the atmosphere was wrung out like a giant wet sponge. Raging floods cascaded down steep mountain slopes and gouged out deep canyons in the rocky plain. When the rains ceased and the skies finally cleared, the Earth was transformed into a giant blue orb, covered almost entirely by an ocean nearly 2 miles deep. Only about 5 percent of the present continental crust existed at that time, which suggests that the early Earth was dominated by a nearly global ocean.

This scenario for the creation of the ocean is supported by ancient marine sediments that have been found in the metamorphosed rocks of the Isua formation in southwestern Greenland (FIG. 1-10). Dating back 3.8 billion years, these are among the oldest rocks found on Earth. They support the idea that the planet had surface water at this time. Between the last of the great meteorite bombardments and the formation of sedimentary rocks, the Earth's surface was flooded by vast quantities of water.

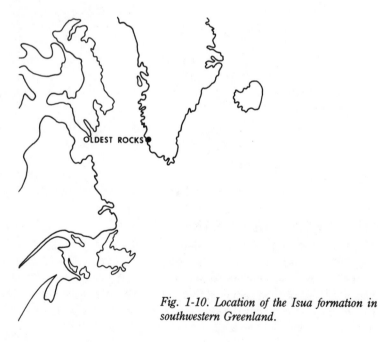

Fig. 1-10. Location of the Isua formation in southwestern Greenland.

Surface temperatures were maintained at high levels due mostly to the greenhouse effect. Sunlight was still about 10 percent less than it is today, but greenhouse gasses, mostly carbon dioxide, methane, and water vapor, held in much of the heat received from the Sun.

Opposing the greenhouse effect were thick clouds of condensed water vapor that completely shrouded the planet much like they do on Venus. The clouds reflected the Sun's rays back into space by what is called the albedo effect (FIG. 1-11). The clouds also kept some of the surface heat from escaping into space by reflecting it back to the ground.

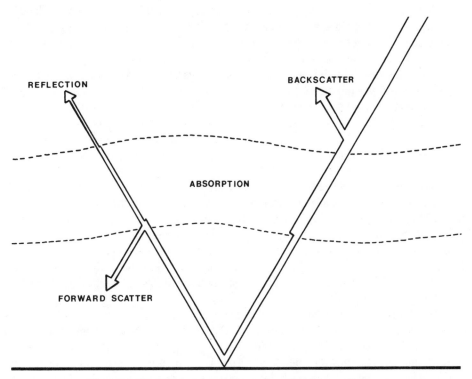

Fig. 1-11. The effect of the albedo on incoming solar radiation.

The climate, which was much warmer then than it is today, supported the evaporation of large quantities of water from the ocean. This created a powerful atmospheric heat engine that pushed strong air currents around the world. Billowing clouds, formed by water vapor, were pulled along by these swift winds. Huge circulating weather systems were generated by a strong Coriolis effect, which is the quality of a rapidly rotating sphere that translates into swirling eddies in a fluid medium such as air or water. Hurricanes many times more powerful than they are today prowled the ocean stretches one after another.

Ocean currents, unimpeded by land, evenly distributed the Sun's energy around the world. This, along with greenhouse gasses in the atmosphere that held in much of the

Sun's heat, maintained the Earth at a nearly constant temperature from pole to pole. The ocean was also heated from below by active volcanoes on the ocean floor. This thickened the oceanic crust and continually supplied seawater with life-giving nutrients.

THE OCCURRENCE OF LIFE

Chaos pervaded the universe then and now. Many of the motions of our Solar System are driven by chaos. Life attempts to create order out of chaos, an uphill struggle that comes at the expense of a great deal of energy, which must be obtained from the Sun. This struggle is manifested by the presence of large amounts of oxygen in the Earth's atmosphere. Without life, chemical reactions would have run steadily downhill and oxygen would have long since vanished. Thus, life seems to maintain oxygen and carbon dioxide in perfect balance. Too much of one with respect to the other could spell the difference between life and death on Earth.

Organisms might have developed photosynthesis as early as 3.5 billion years ago. The oxygen that was produced in this manner, however, was quickly used up by chemical reactions that permanently stored it in the crust. Then, about 2 billion years ago, these oxygen traps became full and the gas began to slowly build up in the ocean and atmosphere. In addition to the generation of oxygen, simple plants removed carbon dioxide from the environment and buried it in the Earth's crust in the form of carbonaceous sediments.

About this time, mobile crustal plates on the Earth's surface began to move extensively. This caused carbonaceous sediments and the oceanic crust to be thrust deep into the Earth. The newly formed surface area increased the amount of carbon dioxide stored in thick deposits of carbonate rocks such as limestone. The first microscopic plants also developed at this time and began to replace the carbon dioxide in the atmosphere with oxygen. The combined loss of carbon dioxide caused the climate to cool even though the Sun was becoming progressively hotter. This initiated the first great ice age about 2 billion years ago.

Another substantial carbon dioxide repository was the great coal forests that spread over the land during the Upper Paleozoic era 260 million years ago. Plants invaded the land about 450 million years ago and extended to all parts of the Earth. Lush forests that grew during the Carboniferous period absorbed large quantities of carbon dioxide. Rapid burial under *anaerobic*, or oxygen-lacking, conditions converted the carbon in the vegetation into thick seams of coal. The burning of coal reverses this process, releasing large quantities of carbon dioxide into the atmosphere.

THE WARM CRETACEOUS PERIOD

During the Cretaceous period from about 135 to 65 million years ago, plants and animals were particularly abundant and spread practically from pole to pole. Volcanoes were especially active during this time. They injected massive amounts of carbon dioxide into the atmosphere, which warmed the planet significantly. The Cretaceous period was the warmest period in the Earth's history.

There is no evidence of any permanent ice caps during the warm Cretaceous period. The deep ocean waters, which are now near freezing, were then around 60 degrees

Fahrenheit. The average global surface temperature, which today is about 60 degrees, was about 20 degrees warmer during this time (FIG. 1-12). The temperature difference between the poles and the equator was only about 40 degrees, whereas today that difference is nearly doubled.

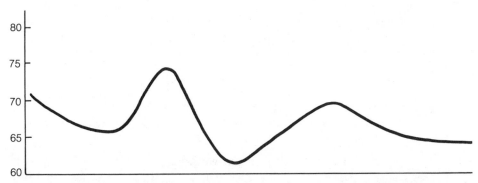

Fig. 1-12. Variation in global temperatures during the Cretaceous period. Temperatures are in degrees Fahrenheit.

The Earth's climate actually began to warm at the beginning of the Jurassic period about 180 million years ago. About this time, the large supercontinent Pangaea began to break up, and the continents drifted into warmer equatorial waters. The oceans were interconnected in the equatorial regions by the Tethys and Central American seaways, which provided a unique current system that completely circled the globe and carried heat toward the poles. The high latitude oceans were less reflective than the land and absorbed more heat, which further moderated the climate.

Coral reefs and other tropical biota, for which bright sunlight and warm seas are essential, ranged as much as a thousand miles closer to the poles than they do today. Polar forests extended into latitudes 85 degrees north and south of the equator. One example is found in the fossilized remains of a forest that once thrived on the now frozen continent of Antarctica. Alligators and crocodiles lived as far north as Labrador, whereas today they are restricted to the warm Gulf Coastal region.

Perhaps the greatest contribution to the warming of the Earth, however, came from increased volcanic activity caused by vigorous continental movements called plate tectonics. Volcanoes produced 4 to 8 times the present amount of atmospheric carbon dioxide (FIG. 1-13), which substantially increased greenhouse warming. This also provided an abundant source of carbon for green plants, which contributed considerably to their prodigious growth and helped feed the hungry dinosaurs.

WHAT KILLED THE DINOSAURS

The dinosaurs were the most successful land animals. They inhabited the Earth for 140 million years. Humans, on the other hand, have only been around for the past 4 million years. The dinosaurs originated during the Triassic period, which began 240 million years ago, when all the landmasses were assembled into the supercontinent Pangaea. During the Jurassic period, about 180 million years ago, the continents split apart.

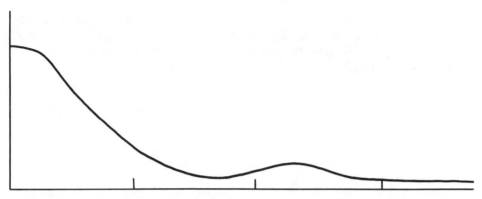

Fig. 1-13. Relative atmospheric carbon dioxide levels over the last 100 million years.

Except for a few temporary land bridges, the newly formed oceans provided a barrier to any further dinosaur migration. At this time, almost identical species lived in North America, Europe, and Africa. The greatest dinosaur that ever lived, brachiosaurus, is found only in Colorado-Utah, southwestern Europe, and eastern Africa. It probably traveled to Africa by way of Europe when the continents were still together.

The success of the dinosaurs is exemplified by their extensive range. They occupied a wide variety of habitats and dominated all other forms of land-dwelling animals. Indeed, if the dinosaurs had not become extinct, mammals would never have achieved dominance over the Earth. Also, humans would not have come into existence because the dinosaurs would have suppressed further advancement of the mammals.

About 500 species of dinosaurs have been discovered thus far, although this is probably only a small fraction of the total. The generally warm climate of the Cretaceous period produced lush vegetation, including ferns and cycads, that supplied the insatiable diets of the plant-eating dinosaurs.

Many theories have been put forward to explain the demise of the dinosaurs (FIG. 1-14). At the end of the Cretaceous period the dinosaurs and 70 percent of all known species vanished. This indicates that something in the environment made them all unfit to survive, yet did not adversely affect the majority of the mammals. The mammals, which were no larger than rodents, coexisted with the dinosaurs for more than 100 million years. They lived a nocturnal life style, however, so as not to compete directly with the dinosaurs.

At the end of the Cretaceous period, the Earth might have been bombarded by a massive meteorite shower. A thin layer of mud at the boundary between the Cretaceous and Tertiary periods provides evidence of such an occurrence. The mud is found in many parts of the world with the greatest concentration in central North America. Within this layer is a high concentration of iridium, an isotope of platinum, and meteoritic amino acids, both of which are extremely rare on Earth, but relatively abundant in meteorites. The iridium level in the layer of mud was a thousand times greater than normal background concentrations, which indicates that meteorite impacts lofted tremendous amounts of dust into the atmosphere and shaded the Earth. This might have cooled the climate enough to cause the extinction of large numbers of species.

Fig. 1-14. A dinosaur boneyard at the Howe Ranch quarry near Cloverly, Wyoming.

On the other hand, the impacts could have caused widespread extinction of microscopic marine plants called calcareous nannoplankton. These plants produce a sulfur compound that when released into the atmosphere helps to make clouds. Clouds, in turn, reflect sunlight and prevent solar radiation from reaching the surface. The death of the calcareous nannoplankton might have triggered an extreme global heat wave that would have killed off the dinosaurs and most other species. Evidence indicates that ocean temperatures did increase dramatically for tens of thousands of years beyond the end of the Cretaceous period.

The breakup of Pangaea might also have contributed to the demise of the dinosaurs by changing global climate patterns and producing unstable weather conditions. Massive floods of basalt from perhaps the most volcanically active period since the Earth's beginning might have dealt a major blow to the climatic and ecological stability of the planet.

Another theory is that a massive bombardment of meteorites could have stripped away the ozone layer and bathed the Earth with the Sun's deadly ultraviolet rays. This would have killed land plants and animals and the primary producers in the surface waters of the ocean. Since the mammals were mostly nocturnal and remained in their underground borrows during the day, they would have survived the onslaught of ultraviolet radiation. This scenario has important implications for us today. For if we continue to destroy the ozone layer with our pollutants, we might find ourselves going the same way as the dinosaurs.

2

The Rock Cycle

ONE of nature's most important cycles is called the geochemical carbon cycle, or simply, the rock cycle. It is unique to the Earth and generated by the planet itself. The rock cycle should not be confused with the biological carbon cycle, which deals with the decay of humus. Presently, carbon dioxide makes up about 0.035 percent of the atmosphere, which amounts to about 700 billion tons of carbon. Carbon dioxide is the primary source of carbon used in the process of photosynthesis by green plants. As such, it provides the basis for all life.

Carbon dioxide traps heat that would otherwise escape into space and acts something like a thermal blanket. In this respect, it plays an important role in regulating the temperature of the Earth. If more carbon dioxide is taken out of the atmosphere than is replenished, the Earth cools down. If more carbon dioxide is generated than is taken out of the atmosphere, the Earth heats up. Therefore, carbon dioxide in the atmosphere functions as a sort of thermostat for the Earth. Major changes in the carbon cycle would have profound effects on the climate.

THE DYNAMIC EARTH

Not even the largest and most prominent features on the Earth's surface, such as mountains and seas, can be regarded as permanent and immovable. The Earth's crust is comprised of rigid lithospheric plates (FIG. 2-1). Because they are constantly in motion, continents and oceans are continuously reshaped and rearranged. Plate tectonics and continental drift were operating early in the Earth's history, possibly as long ago as 2.7

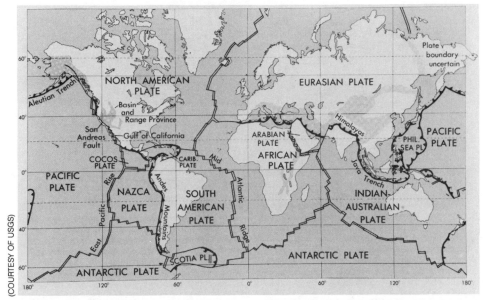

(COURTESY OF USGS)

Fig. 2-1. The Earth's crust is fashioned out of several lithospheric plates whose motions are responsible for all geologic activity.

billion years. Changes in continental configurations greatly affected global temperatures, ocean currents, biological productivity, and many other factors of fundamental importance to the Earth.

The positioning of the continents helped determine climatic conditions. When most of the land huddled near the equatorial regions, the climate was warm. When lands wandered into the polar regions, much of the world became covered with ice. During times of highly active continental movements, there is greater volcanic activity, especially at subduction zones and midocean spreading centers. The amount of volcanism could have affected the composition of the atmosphere and the rate of mountain building, which too can affect the climate.

All the lands were welded into a single large continent called Pangaea, around 280 million years ago. The supercontinent was located near the tropics. This allowed the oceans that existed in the high latitudes, which were less reflective than the land, to absorb more heat. This had a strong moderating influence on the climate. Also, because there was no land in the polar regions to interfere with the movement of warm ocean currents, both poles remained ice free year round. As a consequence, there was no large variation in temperature between the high latitudes and the tropics.

During the initial breakup of Pangaea 180 million years ago (FIG. 2-2), the climate, particularly during the Cretaceous period, was extremely warm. Global average temperatures were 10 to 25 degrees Fahrenheit warmer than they are today. When the continents drifted closer to the poles at the beginning of the Cenozoic era, the land disrupted poleward oceanic heat movement and substituted reflective, easily chilled land for the absorptive, heat-retaining water. As the cooling progressed, the land accumulated snow and ice. This created a much greater reflective surface. The additional loss of solar energy further lowered global temperatures.

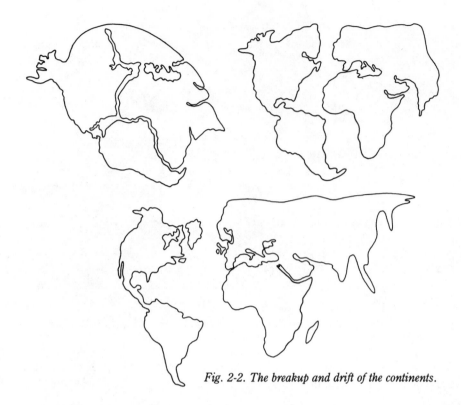

Fig. 2-2. The breakup and drift of the continents.

The Earth's rock cycle has a period of about 300 million years. It is controlled by the cycle of convection currents in the mantle. Convection is the motion that occurs within a fluid medium and results from a difference in temperature from top to bottom (FIG. 2-3).

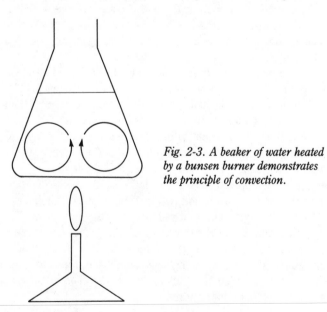

Fig. 2-3. A beaker of water heated by a bunsen burner demonstrates the principle of convection.

Fluid rocks in the mantle receive heat from the core, ascend, dissipate their heat to the lithosphere, cool, and then descend to the core again to pick up more heat. The cycling of heat within the mantle is the main driving force behind all tectonic activity and all other activities that take place on the surface of the planet.

The first phase of the rock cycle was a time of low mantle convection. As a result, the continents were assembled into a supercontinent, the ocean basins widened and the level of the sea dropped, which exposed more land surface. In addition, the level of atmospheric carbon dioxide was reduced due to a decrease in the number of volcanic eruptions. An icehouse effect developed with colder temperatures worldwide. These conditions prevailed from about 700 to 550 million years ago, from about 400 to 250 million years ago, and during the latter part of the Cenozoic era, the period in which we are presently living.

During the second phase of the rock cycle, rapid convection in the mantle lead to the breakup of the supercontinent (TABLE 2-1). This in turn compressed the ocean basins and caused the sea level to rise and flood the land. It also increased volcanism. Vast amounts of lava flooded onto the continental crust during the early stages of many rifts. The increased volcanism also increased the carbon dioxide content of the atmosphere. This resulted in a strong greenhouse effect that produced warm conditions worldwide. These episodes occurred from about 500 to 350 million years ago and again from about 200 to 50 million years ago.

PLATE TECTONICS

All aspects of the Earth's history and structure were fashioned by the mobile crustal plates. The Earth's outer shell is comprised of about a dozen or so plates that are about 60 miles thick. The plates are composed of the lithosphere, which is the rigid outer layer of the mantle, and the overlying continental or oceanic crust. The oceanic crust is remarkable for its consistent thickness and temperature. It averages 4 miles thick and does not vary in temperature by more than 20 degrees Celsius over most of the globe.

The plate boundaries are the midocean ridges, where new oceanic crust is created, and the deep ocean trenches, where the crust is subducted into the mantle and destroyed. No two midocean ridges and no two subduction zones are exactly alike due to the different rates of plate motion. The plates carry the continents with them as they ride on the asthenosphere, which is the semimolten portion of the upper mantle. When two plates collide, the thrust of one under the other uplifts the crust into mountain ranges or creates long chains of volcanic islands. The breakup of a plate creates new continents and oceans. The process of rifting and patching the continents has been going on for the past 2.7 billion years and possibly longer.

Presently, the Atlantic basin is widening and spreading North America and Europe, as well as South America and Africa, apart at a rate of about an inch per year. The continents might have parted faster during the early breakup of Pangaea due to more vigorous plate motions. Average plate motions are observed by satellite laser ranging between the North American, Pacific, and Australian plates. Another method of measurement relies on what is called very-long-baseline interferometry (VLBI). This method uses radio signals from quasars, which are rapidly spinning stars that emit irregular pulses of radio waves. The signals are received at slightly different times by radio telescopes located on

TABLE 2-1. The Drifting of the Continents

AGE IN MILLIONS OF YEARS		GONDWANALAND	LAURASIA
Quaternary period	3		Opening of Gulf of California
Pliocene epoch	11	Begin spreading near Galapagos Islands Opening of the Gulf of Aden	Change spreading directions in eastern Pacific Ocean Birth of Iceland
Miocene epoch	25	Opening of Red Sea	
Oligocene epoch	40	Collision of India and Eurasia	Begin spreading in Arctic Basin
Eocene epoch	60		Separation of Greenland from Norway
Paleocene epoch	65	Separation of Australia from Antarctica	
		Separation of New Zealand from Antarctica Separation of Africa from Madgascar and South America	Opening of the Labrador Sea Opening of the bay of Biscay Major rifting of North America from Eurasia
Cretaceous period	135	Separation of Africa India, Australia, New Zealand, and Antarctica	
Jurassic period	180		Begin separation of North America from Africa
Triassic period	230		
Permian period	280		

two plates. These methods of measurement appear to be in general agreement with the geological rates averaged over millions of years.

The geological rates were determined by paleomagnetic studies of basalts that were laid down on the ocean floor on either side of a spreading ridge. As lava, which is normally iron-rich, extrudes from a rift system and cools, the magnetic fields of its iron atoms align with the Earth's magnetic field like tiny fossil compasses. The geomagnetic field has reversed itself several hundred times over the past 200 million years.

To measure the geological rate, ships traverse the ridge system while towing sensitive magnetic recording instruments. This produces a map of magnetic stripes of varying width and magnitude. The map shows an identical succession of reversing magnetic fields on either side of the ridge (FIG. 2-4). The only reasonable explanation for the identical parallel bands of magnetic rocks on both sides of the midocean ridge is that the ocean floor had to have spread apart at the ridge crests.

How did the continents break apart in the first place? The lithosphere, which includes the crust and its underlying plate, is generally between 50 and 100 miles thick on the continents. On the ocean floor, however, it ranges in thickness from about 5 miles near spreading centers to no more than 60 miles at plate margins. Cracking open the continental lithosphere would appear to be extremely difficult, considering its great thickness.

The best example of the rifting of continents can be found in the East African Rift system, which has not yet fully ruptured. When it does, the present continental rift will be replaced by an oceanic rift. This type of rift is presently taking place in the Red Sea, where Africa and Saudi Arabia are moving away from each other (FIG. 2-5). This transition is accompanied by block faulting. Blocks of continental crust drop down along extensional faults where the crust is being pulled apart. This results in a deep rift valley and a thinning of the crust. In addition, upwelling of molten rock from the mantle further weakens the crust and convection currents pull the crust apart. Therefore, the engine that drives the birth and development of rifts, and the breakup of continents and the formation of ocean basins, ultimately comes from the Earth's interior.

THE SPREADING RIDGES

As rocks heat up in the asthenosphere, they become plastic and slowly rise by convection. After millions of years, the heated rocks reach the underside of the lithosphere. There the rocks spread out in lateral currents and cool, before descending back into the deep interior of the Earth. The constant pressure against the bottom of the lithosphere causes fractures to form. As the convection currents flow out on either side of the crack, they carry the two separated parts of the lithosphere with them, causing the fracture to widen.

The fracture reduces pressure and allows the rocks to melt and rise up through the crack. The molten rock, or magma, passes through the 60 miles or so of lithosphere until it reaches the oceanic crust. There it forms magma chambers that press sideways against the oceanic crust and widen it. At the same time, molten lava pours out from the trough between the two ridge crests and adds layer upon layer to both sides of the spreading ridge. This generates about 3 cubic miles of new oceanic crust every year. The pressure of the upwelling magma forces the ridge further apart and pushes the ocean floor and the lithosphere upon which it rides away from the midocean ridge.

Fig. 2-4a. A crewmember lowers a magnetometer over the stern of the oceanographic research ship USNS Hayes.

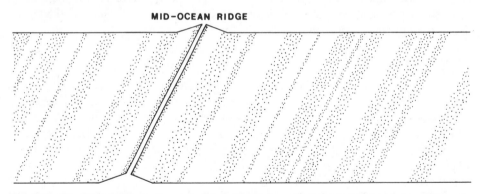

MID-OCEAN RIDGE

Fig. 2-4b. Magnetic stripes on the ocean floor indicate magnetic polar reversal and are evidence for sea-floor spreading.

A huge submarine mountain range called the Mid-Atlantic Ridge runs through the middle of the Atlantic Ocean. It surpasses in scale the Alps and Himalayas combined. What is remarkable about this undersea ridge system is that it bisects the Atlantic Ocean almost exactly down the middle. It weaves halfway between the continents and assumes the shapes of the continental margins on opposite shorelines. The Mid-Atlantic Ridge is the most peculiar mountain range ever known. Down the middle of the 10 thousand-foot-high ridge crest runs a deep trough that looks like it is a giant crack in the Earth's crust. It approaches 4 miles in depth, which is 4 times deeper than the Grand Canyon. The trough is 15 miles wide, making it the longest and deepest canyon on the planet.

The Mid-Atlantic Ridge is part of a worldwide midocean ridge system that stretches over 40,000 miles (FIG. 2-6). The axis of the midocean ridge is offset laterally in a roughly east-west direction. Pieces of oceanic crust slide past each other at the offset points, which are called *transform faults*. The faults range from a few miles to a few hundred miles long and are encountered every 20 to 60 miles along the length of the ridge. Since rocks do not stretch very easily, the crust cracks open, and the pieces of oceanic crust are pushed along by the expansion of the midocean ridge.

Fig. 2-5. The rifting of the Red Sea, viewed from Gemini spacecraft.

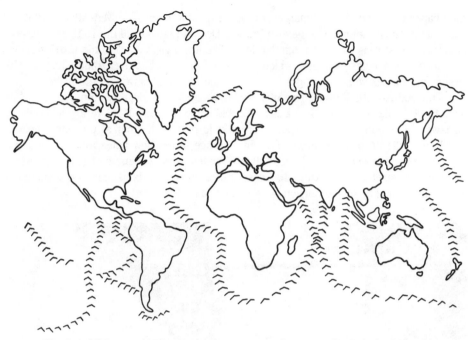

Fig. 2-6. Midocean ridges comprise the most extensive mountain chain in the world.

The friction between the plates gives rise to strong shearing forces that wrench the ocean floor into steep canyons. The grinding of the plates past each other at the transform faults produces strong earthquakes. The crest of the midocean ridge is also a center of intense seismic and volcanic activity and the focus of a great flow of heat from the interior of the Earth. This is the engine that drives all tectonic activity on the face of the planet and is responsible in large part for maintaining the balance of nature. Without plate tectonics, the Earth would become as dead as Mars (FIG. 2-7).

Fig. 2-7. The surface of Mars viewed from Viking Lander 2.

THE SUBDUCTION ZONES

The oceanic crust becomes colder and denser the farther it is from the spreading ridge system. Eventually it is heavy enough to sink into the mantle, at which point it forms a deep-sea trench. The subduction of the crust and lithosphere plays a significant role in global tectonics and accounts for many of the geological processes that shape the surface of the planet. The seaward boundaries of subduction zones are marked by the deepest trenches in the world. These trenches are usually found at the edges of continents or along volcanic island chains (FIG. 2-8). Major mountain ranges and most volcanoes and earthquakes are associated with the subduction of crustal plates.

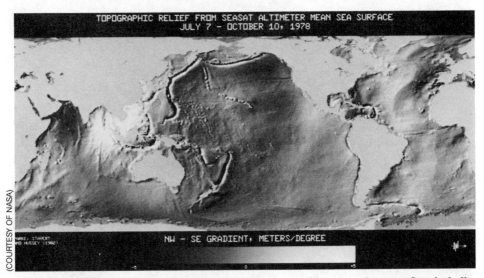

Fig. 2-8. Topographic relief map of the ocean surface showing features on the ocean floor including midocean ridges and trenches.

The amount of subducted plate material is vast. When the Atlantic and Indian oceans opened up and new oceanic crust was created some 125 million years ago, an equal area had to disappear into the mantle. This required the destruction of 5 billion cubic miles of crustal and lithospheric material. At the present rate of subduction, an area equal to the entire surface of the planet will be consumed by the mantle in the next 160 million years.

When two lithospheric plates converge, it is usually the oceanic plate that is bent and pushed under the thicker, more stable continental plate. The line of initial subduction is marked by a deep ocean trench. At first the angle of descent is low. Then it gradually steepens to about 45 degrees. At this angle, the rate of vertical descent is less than that of the horizontal motion of the plate, which is typically 2 to 3 inches per year.

The greater buoyancy of the continental crust prevents it from being carried down into the trench. When two continental plates converge, the crust is scraped off the subducting plate and plastered onto the overriding plate. This welds the two pieces of continental crust together, while the subducted lithospheric plate, now without its overlying crust, continues to dive into the mantle. As additional continental crust is thrusted together, mountain ranges form. This is the process that created the Himalayas (FIG. 2-9) when India collided with Asia about 40 million years ago.

(COURTESY OF NASA)

Fig. 2-9. The Himalayas of India and China viewed from the space shuttle.

Resisting forces will eventually stop the convergence of the plate. Continental collisions, such as the one that happened when India broke off from Gondwana and slammed into Asia, might also be responsible for the periodic reorientation of the plates. These collisions might also explain why the Pacific undersea volcanoes, like the Emperor Seamounts northwest of the Hawaiian Islands, made an abrupt turn to the north about the same time the North American plate rammed into the Pacific plate.

The deep ocean trenches that are created by the descending plates accumulate large deposits of sediments from the adjacent continent. The continental shelf and slope, which consist of sediments washed off the continents and the remains of dead marine life, might extend several hundred miles from the edge of the continent. When these sediments and their content of seawater are caught between a subducting oceanic plate and an overriding continental plate, they are subjected to strong deformation, shearing, heating, and metamorphism, which is the recrystallization of a rock without melting.

Some of the sediments might be carried deep into the mantle, where they melt in pockets called diapirs. Like huge bubbles, the diapirs rise to the base of the continental crust and become a source of new magma for volcanoes (FIG. 2-10). In this manner, the continental crust is rejuvenated and the total mass of low-density crustal rocks always remains the same.

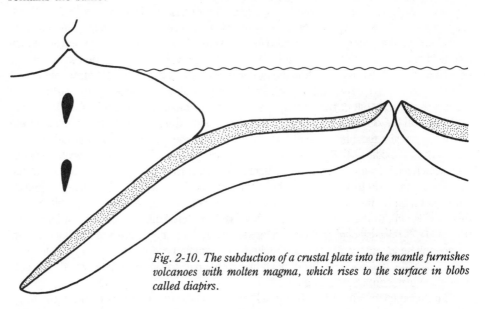

Fig. 2-10. The subduction of a crustal plate into the mantle furnishes volcanoes with molten magma, which rises to the surface in blobs called diapirs.

CARBON RECYCLING

The oceans play a major role in the reduction of the level of atmospheric carbon dioxide. In the upper layers of the ocean, the concentration of gasses is in equilibrium with the atmosphere at all times. These gasses dissolve into the waters of the ocean, mainly by the agitation of surface waves. If the ocean were lifeless, much of its reservoir of dissolved carbon dioxide would enter the atmosphere and more than triple the present level of atmospheric carbon dioxide.

Luckily, the ocean is teeming with life. Marine organisms use carbon dioxide in the form of dissolved bicarbonates to build their carbonate skeletons and other supporting structures. When the organisms die, their skeletons sink to the bottom of the ocean. There they dissolve in the deep waters of the abyssal, which holds by far the largest reservoir of carbon dioxide.

In shallow water, the carbonate skeletons form deposits of limestone, dolostone, or chalk, which bury carbon dioxide in the crust (FIG. 2-11). The burial of carbonate in this manner is responsible for about 80 percent of the carbon deposited on the ocean floor. The rest of the carbonate comes from the burial of dead organic matter that is washed off the continents.

Half of the carbonate in the oceans is transformed back into carbon dioxide, which eventually escapes into the atmosphere. If it were not for this process, in a mere 10,000 years all the carbon dioxide would be taken out of the atmosphere, photosynthesis would stop, and all life would cease. The climate would also turn cold enough to bring on an ice age.

Life in the ocean removes carbon dioxide from the surface waters and the atmosphere and stores it in the sea. The greater the rate of biological activity, the more carbon dioxide is removed from the atmosphere. The rate of removal is determined by the amount of nutrients in the ocean, which corresponds to changes in oceanic ice volume. When the great ice sheets began to melt near the end of the last ice age about 16,000 years ago, the level of carbon dioxide in the atmosphere began to rapidly increase up until about 10,000 years ago. The influx of meltwater flooded continental shelves, which removed organic carbon and nutrients. When nutrients were reduced, the biological rate slowed down and deep-sea carbon dioxide was allowed to return to the atmosphere.

The ocean, by virtue of its large mass, contains about 60 times more carbon than the atmosphere. Most of the carbon is stored in carbonaceous sediments on the continents and on the ocean floor. The amount of carbon in the form of carbon dioxide in the Earth's original atmosphere might have been 1,000 times greater than it is today.

Carbon dioxide in the atmosphere reacts with rainwater to form a weak carbonic acid, which leaches minerals such as calcium and silica from surface rocks. Rivers transport these minerals to the ocean, where they mix with seawater. The calcium minerals are then taken up by marine organisms to make their shells. When the organisms die, their shells sink to the ocean bottom, where they slowly build up deposits of limestone (FIG. 2-11). If this scavenging of carbon dioxide from the atmosphere and storing it in carbonaceous sediments continued unchecked, the atmosphere would become depleted of carbon dioxide, and without this important greenhouse gas, the climate would turn cold enough to bring on a new ice age.

The mystery of what happened to all the carbon dioxide that was found in the original atmosphere was finally solved when the theory of plate tectonics was developed. The ocean floor is continually being created at midocean ridges and destroyed in deep-sea trenches. When the seafloor is forced into the Earth's interior, carbon dioxide is driven out of the limestone by the intense heat. It works its way up through the mantle and eventually ends up in the magma chambers of volcanoes and midocean ridges. The eruption of volcanoes and the flow of molten rock from midocean ridges resupplies the atmosphere with new carbon dioxide, making the Earth one great carbon dioxide recycling plant (FIG. 2-12).

Carbon dioxide plays an important role in regulating the temperature of the Earth. Major changes in the carbon cycle would have had profound effects on the climate. As the early Sun heated up and the temperatures on Earth rose, more water evaporated from the oceans, which increased the amount of rainfall on the land. This speeded up the removal of atmospheric carbon dioxide, which was converted to limestone on the seafloor.

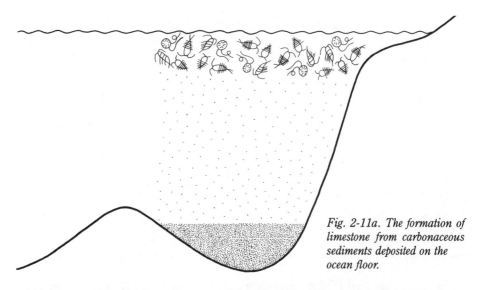

Fig. 2-11a. The formation of limestone from carbonaceous sediments deposited on the ocean floor.

Fig. 2-11b. Intensely folded limestone formation in Atacama Province, Chile.

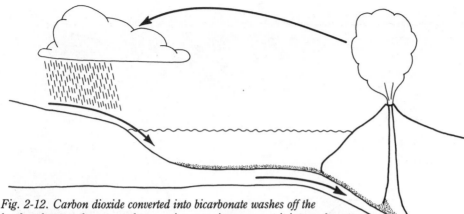

Fig. 2-12. Carbon dioxide converted into bicarbonate washes off the land and enters the ocean where marine organisms convert it into carbonate sediments, which are thrust into the Earth's interior and become part of the molten magma. The carbon dioxide escapes into the atmosphere by volcanic eruptions.

The drop in the level of atmospheric carbon dioxide kept the Earth from overheating. As the Earth's temperature began to cool, less water was evaporated from the ocean. Chemical and biological reactions also slowed down so that less carbon dioxide was removed from the atmosphere, even though the input from volcanoes and spreading ridges remain constant. Thus, the carbon cycle keeps global temperatures within tolerable limits for life.

VOLCANIC ERUPTIONS

Volcanic eruptions and lava flows at midocean ridges are the final stage of the rock cycle. Volcanoes also play a direct role in the Earth's climate. Large volcanic eruptions spew massive quantities of ash and aerosols into the atmosphere where they block out sunlight. Volcanic dust also absorbs solar radiation. This heats the atmosphere and causes thermal disturbances and unusual weather. Scientists were able to date the massive volcanic eruption that destroyed the Aegean island of Thera in 1628 B.C. by studying the stunted growth rings of ancient oaks that are buried in the peat bogs of Ireland, and by looking for signs of atmospheric volcanic ash in the Greenland ice sheet.

Some 400 active volcanoes, called the ring of fire, surround the Pacific Ocean. Indonesia's explosive volcanoes (FIG. 2-13) have produced more violent blasts in historic times than any other region. These volcanoes are associated with subduction zones along the rim of the Pacific basin.

Rift volcanoes account for about 15 percent of the world's known active volcanoes. Most are in Iceland and eastern Africa. It is estimated that there are also about 20 eruptions of deep submarine rift volcanoes every year. Rift volcanoes on continents such as those in eastern Africa can be highly explosive.

Hot spot volcanoes, which are usually found in the interior of crustal plates, are the result of plumes of hot material that rise to the surface from deep within the mantle. The most prominent hot spot volcanoes created the Hawaiian Islands (FIG. 2-14). The newest volcano, Loihi, lies to the southeast of the main island of Hawaii. It is presently building an undersea mountain that will someday break the surface of the ocean and become a new island.

(COURTESY OF NASA)

Fig. 2-13. An active volcano on Andonara Island, Indonesia.

Volcanic activity plays an important role in the restoration of atmospheric carbon dioxide. Carbon dioxide is one of the most important volatiles in magma because it helps make lava flow easily. Carbon dioxide escapes from carbonaceous sediments when they melt after being forced into the mantle at subduction zones near the edges of crustal plates. The molten magma, along with its content of carbon dioxide, rises to the surface to feed volcanoes that lie on the edge of subduction zones and at midocean ridges. When the volcanoes erupt, carbon dioxide is released from the magma and returned to the atmosphere, completing the cycle.

(COURTESY OF USGS)

Fig. 2-14. The broad shield volcano Mauna Loa built most of the island of Hawaii.

The Water Cycle

THE movement of water over the Earth is one of nature's most important cycles. If water did not evaporate from the ocean, fall as rain on the land, and then return to the ocean, there would be no life as we know it. The average journey of a water molecule from the ocean, to the atmosphere, to the land, and back to the sea again takes about 10 days. The journey is only a few hours long in the tropical, coastal regions, but might take as much as 10,000 years in the polar regions. This is what is known as the hydrological, or water, cycle (FIG. 3-1).

The quickest route water can take back to the ocean is through runoff into streams and rivers. This is perhaps the most important part of the water cycle. Rivers provide waterways for commerce and water for irrigation, hydroelectrical power, and recreation. Surface water runoff supplies minerals and nutrients to the ocean and cleanses the Earth. The importance of water to life is so obvious that it is too often overlooked. Also, much of the water has become polluted by human activities, and the accumulation of toxic substances in the ocean could cause substantial damage to the biosphere that might be irreparable.

THE ENERGY BUDGET

The atmosphere plays an important role in maintaining a balance between incoming solar radiation and outgoing infrared radiation. If the Earth had no atmosphere to distribute the Sun's heat over its surface, average global temperatures would be near 0 degrees, or about 60 degrees Fahrenheit colder than they are today. The Earth intercepts only about a billionth of the Sun's rays. Of those rays, only half reach the surface and 90 percent of those that do, evaporate seawater to make the weather.

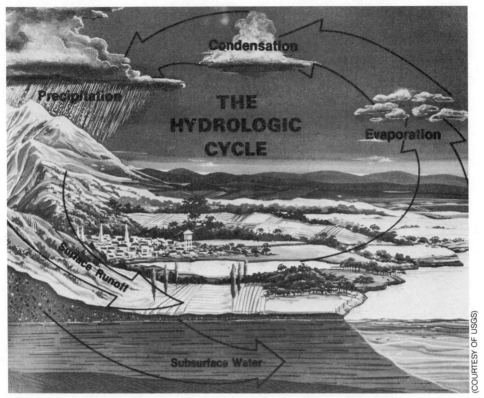

Fig. 3-1. The hydrologic cycle is a continuous flow of water from the ocean, over the land, and back to the sea.

When sunlight strikes the Earth's surface, it is converted into infrared energy. This energy is absorbed by the atmosphere and emitted back to space from levels between 15,000 and 20,000 feet in altitude. The Earth must reradiate back into space exactly the same amount of energy it receives from the Sun or it will become intolerably hot. If the Earth emits too much infrared energy, however, it will become exceedingly cold. This delicate balancing act is known as the Earth's energy or heat budget (FIG. 3-2) and is responsible for maintaining the Earth's temperature within a narrow range that makes life possible.

In the tropics, the Sun's rays strike the Earth from directly overhead. This causes more solar radiation to be absorbed at the surface than is reflected back into space. In the polar regions, the Sun's rays strike the Earth at a shallow angle, so more solar radiation is reflected back into space than is absorbed on the surface. If it were not for the distribution of heat by the atmosphere, conditions, on Earth would be unlivable.

Warm air at the equator rises in narrow columns and travels aloft toward the poles. There it liberates its heat, cools, sinks, and returns to the equator to be warmed again. Currents in the ocean act much the same way, only the process is slower and the journey takes a lot longer. The middle latitudes, or temperate zones, become a battleground between warm, moist tropical air and cold, dry polar air. When these air masses clash, storms develop.

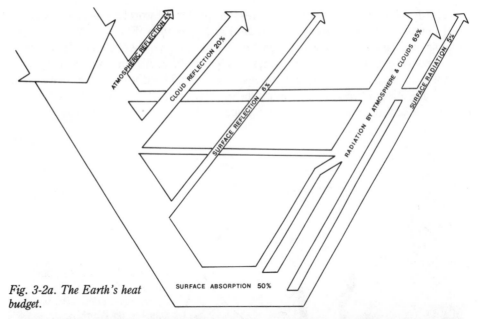

ATMOSPHERIC REFLECTION 6%

CLOUD REFLECTION 20%

SURFACE REFLECTION 6%

RADIATION BY ATMOSPHERE & CLOUDS 65%

SURFACE RADIATION 5%

SURFACE ABSORPTION 50%

Fig. 3-2a. The Earth's heat budget.

Fig. 3-2b. The Earth Radiation Budget Explorer (ERBE) satellite is designed to study the Earth's heating patterns in order to anticipate climate trends that affect agriculture and other resources.

The distribution of air masses is also responsible for the world's winds. The Coriolis effect bends air currents in response to the Earth's rotation (FIG. 3-3). A point on the ground moves faster at the equator than it does near the poles because it is farther away from the axis of rotation and must travel a greater distance. Air currents moving toward the poles are deflected to the east because the ground beneath them is rotating at a slower rate. Air currents moving toward the equator are deflected to the west because the rotation of the ground beneath them is speeding up.

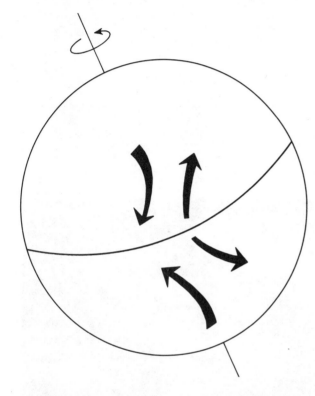

Fig. 3-3. The principle of the Coriolis effect. As the Earth rotates counterclockwise, the winds in the Northern Hemisphere are deflected to the right, while those in the Southern Hemisphere are deflected to the left.

The world's oceans play an important role in distributing solar energy. Heated by strong solar radiation, warm seawater turns into water vapor which makes clouds. When the clouds move to other parts of the world, they radiate energy back to the earth by precipitation, which effectively distributes the ocean's heat around the world (FIG. 3-4).

The oceans are also responsible for the steady onshore and offshore breezes. During the day, the land is warmer than the sea. Warm air rises from the land and travels aloft toward the sea, where it cools and descends back to Earth. At night, the temperature of the land falls below that of the sea. Warm air rises from the sea and travels aloft toward the land, where it cools and again drops to Earth. The monsoon winds operate under similar circumstances except that they are a seasonal phenomena.

The amount of solar energy received on the ground, averaged over a year and spread evenly around the world, is over a million watts for an area equal to the size of a football field. This is about 5,000 times greater than the energy that radiates from the Earth's interior. Because rocks are such effective insulators, very little heat escapes from the interior to warm the surface. The exceptions are hydrothermal vents that are located at volcanic spreading ridges. Here, deep ocean bottom temperatures are warm enough to support prodigious life, which is totally independent of the Sun for its energy (FIG. 3-5).

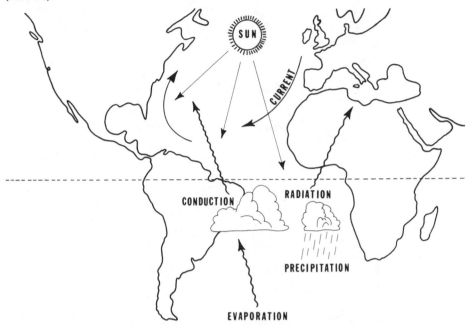

Fig. 3-4. Heat flow between the ocean and the atmosphere.

(COURTESY OF USGS)

Fig. 3-5. Hydrothermal vents on the deep ocean floor provides nourishment and heat for bottom dwellers.

The heat budget is dependent mainly on the Earth's albedo effect. Some objects reflect solar energy better than others (TABLE 3-1). Light-colored objects like clouds, snow fields, and deserts reflect more solar energy than they absorb, while dark-colored objects like oceans and forests absorb more solar energy than they reflect. Most of the solar energy impinging on the ocean is used to evaporate seawater. Ultimately, this energy is given up and lost to space when water vapor condenses into rain. The process is similar to that of an air conditioner, which absorbs heat from inside a house and pumps it outside.

TABLE 3-1. Albedo of Various Surfaces

SURFACE	PERCENT REFLECTED
Clouds, stratus	
< 500 feet thick	25 – 63
500 – 1000 feet thick	45 – 75
1000 – 2000 feet thick	59 – 84
Average all types and thicknesses	50 – 55
Snow, fresh-fallen	80 – 90
Snow, old	45 – 70
White sand	30 – 60
Light soil (or desert)	25 – 30
Concrete	17 – 27
Plowed field, moist	14 – 17
Crops, green	5 – 25
Meadows, green	5 – 10
Forests, green	5 – 10
Dark soil	5 – 15
Road, blacktop	5 – 10
Water, depending upon sun angle	5 – 60

The angle of sunlight incident on the Earth's surface also determines how much solar energy is absorbed and how much is reflected away. In the tropics, the Sun's rays strike the Earth at a high angle, which increases absorption of solar radiation. In the polar regions, the Sun's rays strike the surface at a low angle and solar radiation glances off into space. This, along with a high albedo, helps keep the polar regions permanently ice-bound.

Solar energy is also scattered sideways because of dust particles and aerosols emitted into the atmosphere by volcanic eruptions, dust storms, forest fires, sea salt, meteors, and air pollution. These fine particles are responsible in large part for making the sky blue. If it were not for the dispersion of light by the atmosphere, the daytime sky would be as black as night and the Sun would look as though it were an exceedingly large star. When the Sun is low on the horizon, its rays must pass through so much atmosphere that only the reds get through, which produces beautiful ochre-colored sunrises and sunsets.

About one-third of the Sun's energy is reflected back into space before it ever has a chance to heat the Earth. Most of this lost energy is reflected off the tops of clouds. At any given time, half the skies are covered with clouds (FIG. 3-6), which have a high albedo. The underside of clouds also reflect escaping infrared energy back to the ground, which is why nights are warmer under an overcast sky and colder under a clear one. Different types of clouds have different properties so that low, thick stratocumulus clouds tend to cool the Earth, while high thin cirrus clouds tend to warm it because they function like a greenhouse gas.

(COURTESY OF NASA)

Fig. 3-6. View of the Earth from Apollo 17, *showing extensive cloud formations along with the south polar ice cap.*

On a global average, however, clouds tend to cool the Earth more than they heat it. As the Earth warms, clouds gain cooling power and slow down the heating. The heating and cooling effects of clouds almost balance out over the tropics. At the mid-latitudes, clouds do their most cooling. Aerosols, which are fine solid or liquid particles that have been injected into the atmosphere by natural causes such as volcanoes or by man-made pollutants, also block the Sun's heat from reaching the ground. They allow infrared heat leaving the ground to escape into space, however, which has a net cooling effect.

Only about one-half the total solar energy reaches the surface, where, besides heating the ocean, it is absorbed by soil and plants. The plants use the red and blue rays for photosynthesis, but have no use for the greens, which are reflected away. This is why most plants are green. Eventually, all the sunlight that manages to reach the surface of the Earth is converted into infrared energy and is radiated out to space. If it were not for the greenhouse effect, which prevents all the infrared energy from departing the Earth, the planet would indeed become a very cold place to live.

THE STATES OF WATER

The Earth is the only body in the Solar System where water is known to exist in all three states: solid, liquid, and gas. Each water molecule is composed of an oxygen atom and two hydrogen atoms, separated by electrical charges at an angle of about 105 degrees. The hydrogen atoms have a slight positive charge, and the oxygen atom has a weak negative charge. These charges allow water molecules to clump together and form groups of up to eight molecules (FIG. 3-7).

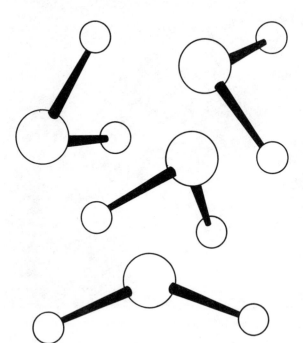

Fig. 3-7. As the temperature nears freezing, water molecules group together.

More groups of water molecules form when temperatures near the freezing point. Because these groups take up more space, water expands upon freezing, which is just the opposite of what most natural materials do. Ice is therefore less dense and able to float on water. This is fortunate because if ice were allowed to sink to the bottom of the ocean where the temperature is already near freezing, it might accumulate to such an extent that the entire ocean would become a solid block of ice.

It requires 80 calories of heat to melt one gram (0.035 ounce) of ice. This is known as the *latent heat of melting*. It takes 100 calories of heat to raise the temperature of one gram of ice to the boiling point, which is called *sensible heat*. And 540 calories of heat is needed to turn water into water vapor, which is known as the *latent heat of vaporization* (FIG. 3-8). The conversion of water vapor back to one gram of ice liberates a total of 720 calories of heat. This is a lot of energy. Water has the highest specific heat, or heat capacity of any natural substance. Water is also an excellent solvent. Most of the chemical elements on Earth, including some man-made chemicals, are able to dissolve in water.

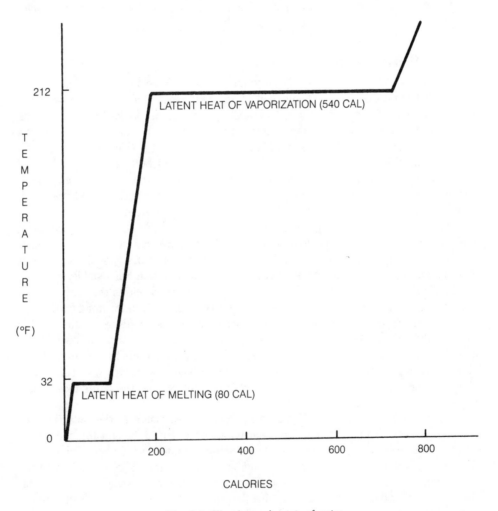

Fig. 3-8. The change in state of water.

Three percent of the Earth's water is fresh. There is enough fresh water to fill the Mediterranean Sea 10 times over. Roughly 25,000 cubic miles of fresh water showers down on the land annually. About three-quarters of all the fresh water is locked up in glacial ice, 90 percent of which exists in Antarctica. There is nearly 6 million square miles of ice on the frozen continent at an average thickness of about 1.3 miles.

Less than 1 percent of the Earth's water is in atmospheric water vapor, rivers, lakes, groundwater, soil moisture, and plant and animal tissues. At its present temperature, the atmosphere can only hold about one-half of 1 percent of the Earth's water at any given time. However, during warmer periods, the atmosphere can hold much higher levels of water vapor, which is the most effective greenhouse gas.

About 15 percent, or some 15,000 cubic miles, of the moisture in the atmosphere is supplied by land. The majority of the water that falls on land eventually makes its way back to the sea. This completes the final and most important loop of the hydrologic cycle.

OCEAN CURRENTS

The ocean's ability to effectively store and transport vast quantities of heat has a profound effect on the climate. The ocean's large heat capacity allows it to retain summer's heat and release it in winter. This ability enables the ocean to moderate the Earth's temperature throughout the seasons. Every summer the ocean surface warms by as much as 25 degrees Fahrenheit above its preceding winter value. It takes about a decade to significantly change the temperature of the upper 1,000 feet of the ocean and thousands of years to significantly change the temperature of the entire ocean. This is what is called the oceanic thermal lag. The heat capacity of the ocean is so large that it might take tens to hundreds of years for it to respond fully to greenhouse warming.

Around 40 million years ago, the warm, salty waters of the ocean's surface might have traded places with its cold deep waters. This would have caused the entire oceanic circulation system to run backwards. Then about 30 million years ago, the continents shifted their positions and the present oceanic circulation system was established.

The surface and abyssal currents in the ocean play a vital role in the movement of heat around the planet (FIG. 3-9). The surface currents are driven by steady wind currents. As with the flowing air masses, surface currents are also deflected by the Coriolis effect to the right in the Northern Hemisphere and to the left in the Southern Hemisphere. Oceanic surface currents are similar in function to currents in the atmosphere. They distribute warm tropical water to the higher latitudes and return with cold water. The difference between air and surface oceanic currents is that the latter move much slower.

Abyssal currents are driven by thermal forces. Cold water in the polar regions descends, spreads out upon hitting the ocean floor, heads toward the equator, and is then distributed in tropical waters. The upwelling of deep seawater in the tropics produces a high concentration of dissolved carbon dioxide over the equator. The path taken by the deep-water currents is influenced by the distribution of landmasses and by the topography of the ocean floor. Abyssal currents that flow toward the equator are deflected to the west because of the Earth's eastward rotation. This presses the currents against the eastern edges of the continents.

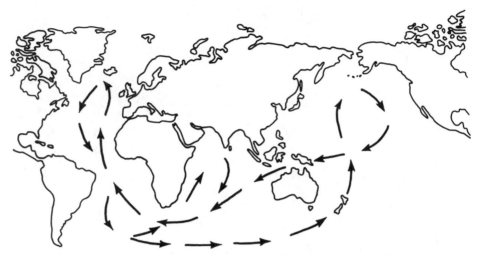

Fig. 3-9. The ocean's heat transport systems distribute warm water to cold regions.

The cold, salty surface waters of the Arctic are dense enough to sink and form a deep-sea current called the North Atlantic Deep Water (NADW). NADW is a subsurface ocean river. Its flow is 20 times greater than the combined flow of all the world's rivers. Another subsurface current, called the western boundary undercurrent, flows along eastern North America and transports some 20,000 cubic miles of water annually.

The sinking water in the poles is matched by upwelling currents in the tropics, creating an efficient heat transport system. It takes upwards of 1,000 years to complete the journey from the tropics to the poles and back again. When the cold water of the abyss reaches the tropics, it rises toward the surface. These upwelling currents play an important role in transporting nutrients from the deep ocean bottom to the surface, where they support marine life. Although these zones cover only about 1 percent of the ocean's surface area, they sustain about 40 percent of all sea life.

Some parts of the oceanic currents get pinched off and form eddies or rings of swirling water like undersea hurricanes. Some of these eddies are as much as 100 miles or more across and reach as deep as 3 miles below the surface, although most are less than 50 miles across. These rings mix the ocean waters. Marine life often get trapped in these rings and are transported to hostile environments. They survive only as long as the rings do, which is usually several months.

Oceanic currents have a dramatic affect on the weather. Changes in these systems can send abnormal weather patterns all around the world. Once every 5 to 8 years, anomalous atmospheric pressure changes in the South Pacific, called an El Niño Southern Oscillation (ENSO), causes the westward-flowing trade winds to collapse (FIG. 3-10). Warm water that is piled up in the western Pacific Ocean by these winds flows back to the east. This creates a great sloshing of water in the South Pacific basin. The layer of warm water in the eastern Pacific Ocean becomes thicker and suppresses the thermocline, which is the boundary between cold and warm water layers. Cold water is prevented from rising. This temporarily disrupts the upwelling of nutrients, which adversely affects local marine biology.

The opposite condition results when the surface waters of the Pacific Ocean cool

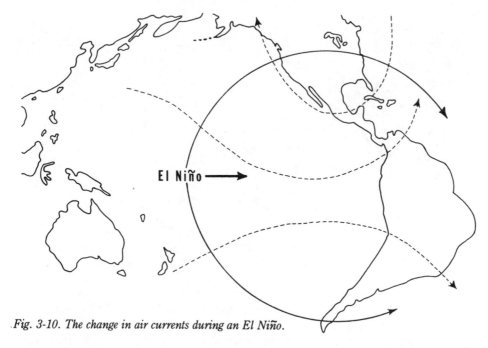

El Niño →

Fig. 3-10. The change in air currents during an El Niño.

during what is called a La Niña. In mid 1988, water temperatures in the central Pacific Ocean plummeted to abnormally cold levels. This signaled a climate swing from an El Niño to a La Niña. As a result, strong monsoons hit India and Bangladesh and heavy rains fell on Australia that year. The La Niña might also have been responsible for a severe drought in the U.S. during 1988 and a marked drop in global temperatures in 1989.

Waves on the ocean are produced by the wind blowing across the surface of the water. Storms at sea are responsible for most of the waves that strike against nearby shores. Offshore hurricanes can produce the largest waves, some of which can cause extreme damage. The wind-stirred layer of the ocean involves the upper 100 to 200 feet. It is the most uniform environment on Earth and is in equilibrium with the atmosphere at all times. However, it is only a thin film floating on the surface of an over 2-mile-deep ocean, the vast bulk of which is near freezing.

Most of the marine life in the ocean lives in the mixed zone called the phototropic zone. These marine organisms must live near the surface in order to have sunlight for photosynthesis. The surface action of the ocean plays an important role in the exchange of carbon dioxide and oxygen. Eighty percent of the Earth's total supply of oxygen is generated by marine plants. However, if none of the oxygen were removed by respiration and decay, its level could double in about 10,000 years, and the Earth would incinerate itself.

Another transfer mechanism from the sea to the air is the transport of marine-borne substances by the wind. These materials are ejected into the atmosphere by bursting air bubbles and spray from waves. The fine spray evaporates into small particles of sea salt that are sent aloft by air currents. Upwards of 10 billion tons of salt enter the atmosphere in this manner annually. The salt also provides seed crystals for the condensation of rain.

WATER ON THE LAND

Seventy percent of the Earth's surface is covered by oceans, which have an average depth of 3 miles. This is a great deal of water—nearly one-quarter of a billion cubic miles. Everyday, a trillion tons of water rains down on the Earth. Most of this rain, however, falls back into the ocean. An uneven distribution of rainfall on land causes deserts and tropical rain forests, along with droughts and floods.

On average, 25 inches of rain falls annually over the Earth's entire surface. Some places receive more precipitation than others (FIG. 3-11). Tropical rain forests receive upwards of 200 to 400 inches per year, while deserts receive less than 10 inches. The amount of precipitation on land amounts to about 25,000 cubic miles. Approximately 10,000 cubic miles of this rain is surplus water, which is lost through floods or held by soil or swamps. About a third is base flow, or the stable runoff of all the world's streams and rivers. Another third is subsurface flow, which discharges mostly through evaporation. Only about 1 percent of groundwater discharges directly into the sea.

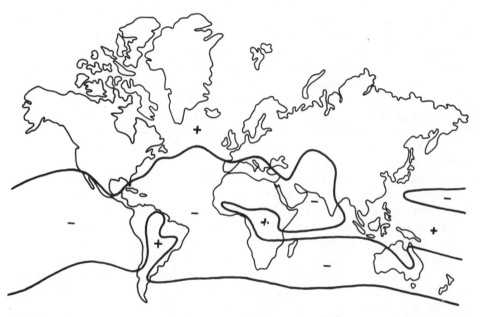

Fig. 3-11. The precipitation-evaporation balance of the Earth. In positive areas, precipitation exceeds evaporation. In negative areas, evaporation exceeds precipitation.

Groundwater is the second most important source of fresh water. It is caught in aquifers that are composed of unconsolidated sand and gravel. Water flows through this formation by the influence of gravity at a very slow rate, perhaps only a few feet a year. Catchment areas at the head of the aquifer recharge the groundwater system. The rate of infiltration into the groundwater system depends on the distribution and amount of precipitation, the type of soils and rocks, the slope of the land, the amount and type of vegetation, and the amount of water rejected because the ground is oversaturated with soil moisture.

The over-use of groundwater can lower the water table or deplete the aquifer altogether. Once an aquifer is depleted, it can no longer be restored to its original capacity because subsidence caused by the weight of the overlying strata compresses the sediments. This in turn decreases the pore spaces between grains, through which the water flows.

There are about 3 million miles of rivers in the United States. About 6 percent of the land area that adjoins these water courses is prone to flooding. Floods are natural reoccurring events and are important because they distribute soil. During a flood, a river might change its course as it meanders its way to the ocean. In so doing, it carves out a new flood plain. Flood plains provide level ground and fertile soil, but they also attract commerce. Rapid development of these areas, without due consideration of the flood hazards, usually ends in disaster with the destruction of property and the loss of a great many lives.

THE FINAL LEG

Acidic rainwater reacts chemically with metallic minerals on the continents. This produces metallic salts that are carried in solution by rivers that empty into the ocean. Rainwater also percolates into the ground, dissolves minerals from porous rocks, and transports these to the ocean by aquifers. Solid rock exposed on the land is chemically broken down into clays and carbonates and mechanically broken down into silts, sands, and gravels. Rivers heavily laden with sediment often fill their beds and are forced to take several different pathways as they head toward the sea (FIG. 3-12). Approximately 15 billion tons of continental material reaches the outlets of rivers and streams annually.

When rivers reach the ocean, their velocity falls off sharply. Their sediment load drops out of suspension and chemical solutions are thoroughly mixed with seawater by currents and wave action. These elements are distributed evenly throughout the ocean with a mixing time of about 1,000 years.

Sediments continually build the continental margins outward. Coarser sediments accumulate near shore and progressively finer sediments settle out of suspension farther away from the shore. As the shoreline moves seaward, the original fine sediments are covered by coarser sediments. As the shoreline recedes because of higher sea levels, coarse sediments are covered by fine sediments. This provides a sedimentary sequence of sandstone, siltstone, and shale. Also, carbonates precipitate and accumulate in thick beds on the ocean floor.

The sediments that end up on the ocean floor are detritus and the shells and skeletons of dead microscopic organisms that flourish and die in the sunlit waters of the mixed layer of the ocean. Detritus is generated by the weathering of surface rocks along with decaying vegetable matter. It is carried by rivers to the edge of the continent and out onto the continental shelf. When the detritus reaches the edge of the continental shelf, it falls to the base of the continental slope (FIG. 3-13) under the pull of gravity. Most of the detritus is trapped near the outlets and on continental shelves. Only a few billion tons actually escape into the deep sea.

Fig. 3-12. India's Brahmaputra River Valley viewed from space.

The biological material in the sea contributes about 3 billion tons of sediment, which accumulates on the ocean floor each year. The rates of accumulation are governed by the rates of biological productivity. These in turn are controlled in large part by the ocean currents. When nutrient-rich water upwells from the ocean depths to the sunlit zone, the nutrients are incorporated into the cells of microorganisms. Areas of high productivity and high rates of accumulation are normally found around major oceanic fronts, such as the region around Antarctica, and along the edges of major currents, such as the Gulf Stream and the Kuroshio or Japan current.

Nutrient-rich water also upwells in an equatorial zone along the coasts of Ecuador and Peru. The rate of marine life sedimentation is also influenced by ocean depth. The farther the shells have to descend, the greater the chances that they will dissolve before reaching the bottom. A shell's survivability also depends on how quickly it is buried by sediments and protected from the corrosive action of the deep-sea water.

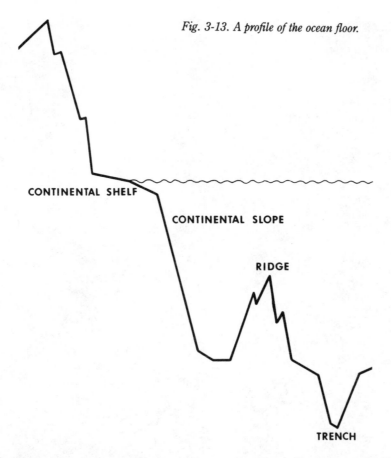

Fig. 3-13. A profile of the ocean floor.

CONTINENTAL SHELF

CONTINENTAL SLOPE

RIDGE

TRENCH

If there were no bottom currents and only marine-born sedimentation, an even blanket of material would settle onto the original volcanic floor of the world's oceans. Instead, the rivers of the world contribute a substantial amount of the material that ends up on the deep ocean floor. The largest rivers of North and South America empty into the Atlantic Ocean, which receives considerably more river-borne sediment than the Pacific Ocean. The Atlantic is smaller and shallower than the Pacific, so its marine sediments are buried more rapidly and are more likely to survive than their counterparts in the Pacific.

The deep ocean trenches around the Pacific Ocean trap much of the material that reaches its western edge, where it is subducted into the mantle. On average, the floor of the Atlantic Ocean receives considerably more sediment, about an inch per 2,500 years, than the floor of the Pacific. Strong near-bottom currents also redistribute sediments in the Atlantic on a larger scale than they do in the Pacific. Undersea storms carry great quantities of sediment, which is dumped in great heaps on the ocean floor when the storm dies out. This makes the bottom of the ocean one of the most dynamic environments on Earth.

4

The Weather Maker

METEROLOGISTS and climatologists are now beginning to fully understand how the intricacies of the atmosphere relate to weather and climate. Supercomputers are routinely used to model the climate. These general circulation models can make weather predictions far into the future or determine what the climate might have been like in the distant past. However, the accuracies of these predictions are dependent on so many variables that they become little more than educated guesses.

Increased temperatures from greenhouse warming could energize the atmosphere to such an extent that storm systems would become much more violent. In addition, instabilities in the atmosphere could change weather patterns so that once productive areas become deserts, while other areas receive severe floods and the resultant soil erosion. There might also be larger seasonal temperature variations and a higher moisture content in the atmosphere. This could produce rainstorms or snowstorms of unprecedented proportions and dry winds of tornadic force that would produce dust storms that could turn day into night (FIG. 4-1). Tornadoes, hailstorms, thunderstorms, and lightning could all increase in frequency and intensity. Numerous immense hurricanes could prowl the oceans and charge headlong into heavily populated coastal areas, which would cause tremendous property damage and a great loss of lives.

ATMOSPHERIC CIRCULATION

Weather is dependent on the interaction of the ocean and the atmosphere. The troposphere, which is where weather takes place, is a blanket of air that extends from the Earth's surface to an altitude of 9 to 12 miles. It contains more than 80 percent of the

Fig. 4-1. Dust storm in Boca County, Colorado, where it was as dark as midnight for over an hour.

total mass of the atmosphere. The troposphere is wider at the equator and narrower at the poles, where the air is colder and denser and therefore more compressed. The most striking feature about the troposphere is that it is in motion at all times. Warm air ascends at the equator, moves toward the poles, clashes with cold fronts, and produces storms (FIG. 4-2).

COLD FRONT   WARM FRONT

Fig. 4-2. The formation of frontal storms.

Tropical cyclones, or hurricanes, act like giant whirling mixers that distribute the Earth's heat and rainfall around the planet. Within recent years, meteorologists have begun to understand that the atmosphere and the ocean act almost like a single fluid, exchanging both heat and gasses as the storms suck moisture from the ocean into their vortexes. However, very little is known about the huge expanses of the ocean, especially in the tropics, where much of the worst weather develops.

Air currents travel from the equator to the poles and back again in cellular structures called Hadley cells. They are named for the eighteenth century English meteorologist George Hadley who developed the theory of convection. When a parcel of air is heated at the equator, it rises to the top of the troposphere, an area called the tropopause. There it is blocked from climbing any further and forced to move toward the poles. Because the Earth is spinning beneath it, however, the parcel of air is bent to the east due to the Coriolis effect. When the parcel of air reaches the polar regions, it cools, sinks, and heads back toward the equator along a path that is bent to the west.

The rotation of the Earth prevents the formation of single, large equator-to-pole cells. Instead, the circulation is accomplished by three separate cells in each hemisphere that transfer heat and cold from one cell to the next. These cells are responsible for the world's major wind belts. From equator to pole these wind belts are: the trade winds, the prevailing westerlies, and the polar easterlies (FIG. 4-3). The interaction between cells is complicated by the distribution of oceans, continents, mountain ranges, deserts, forests, and glaciers.

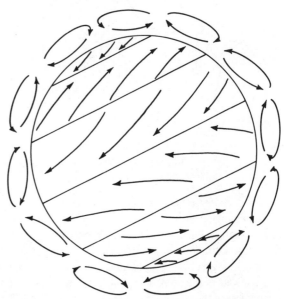

Fig. 4-3. Convection cells in the atmosphere are responsible for distributing heat from the tropics to the poles and for the world's winds.

The greatest amount of heating takes place at the equator where the air is most directly beneath the Sun. The barometric pressure is lower at the equator than it is in the subtropical zones on either side. Therefore, rising air is associated with low pressure and is created when air masses from both sides of the equator rush in, meet head on, and are forced upward. A belt of converging air currents, called the intertropical convergence zone (ITCZ), lies in the equatorial regions. This is where equatorial storms are formed when moisture evaporates from the oceans and is carried toward the equator by the trade winds. The ITCZ varies in width from a few miles to about 60 miles and is usually broader over the oceans. Its position changes from day to day as well as from season to season.

The ITCZ plays a major role in the thermal drive mechanism that creates weather. When warm, moist air rises, it is cooled by expansion because air becomes thinner and colder with altitude. The cooling causes water vapor to condense into clouds. The condensation releases thermal energy when water changes state from gas to liquid. This causes the clouds to rise even higher. With further cooling, precipitation results. The ITCZ is associated with some of the heaviest rainfall areas in the world, including the rain forests and jungles of South America and Asia. As it wanders north and south of the equator, it draws in moisture-laden winds from the Indian Ocean and brings life-giving monsoons to southern Asia, Africa, and Australia (FIG. 4-4).

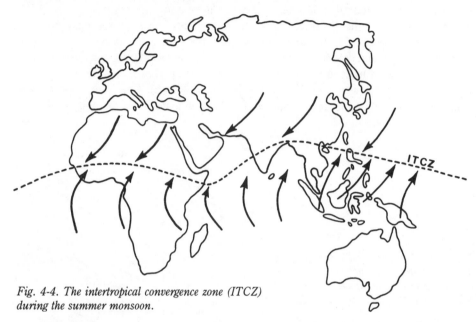

Fig. 4-4. The intertropical convergence zone (ITCZ) during the summer monsoon.

The opposite condition exists in the subtropical zones. There the air sinks, which produces high pressure. After the rain has fallen out of the rising air in the ITCZ, there is little moisture left for the upper air currents to carry to the subtropics. This produces the horse latitudes, which are marked by generally clear skies and calm winds. Most of the world's great deserts lie under the sinking, dry air of the subtropics. As the high-pressure air of the subtropics subsides, it flows back toward the equator to complete the circulation of the tropical Hadley cell. All Hadley cells operate in much the same fashion, with low pressure accompanied by moist, rising air and high pressure accompanied by dry, sinking air.

The primary function of the atmosphere's general circulation pattern is to distribute heat from the tropics to the polar regions. However, if it were not for this circulation of air currents, the tropics would become too hot and the higher latitudes would become too cold to be inhabited. The Earth intercepts about a billionth of the Sun's energy with only about half of the solar energy reaching the surface. On the surface, most of the energy is used to evaporate seawater which forms clouds. When the clouds move to other parts of the world they radiate energy by precipitation, which effectively distributes the ocean's heat.

The general circulation of the atmosphere is also connected with the circulation and temperature of the ocean. The ocean stores heat in the summer and releases it in the winter. The temperature difference from day to night and from summer to winter between the land and the sea is also responsible for the daily onshore and offshore winds and the seasonal monsoons. The ocean's capacity for storing and moving vast quantities of heat surpasses that of the atmosphere by about 40 times. However, if climatic warming were to occur too rapidly, however, the equilibrium between the ocean and the atmosphere could collapse. This would cause a radical change in circulation patterns, which could greatly affect the weather.

The transport of heat by oceanic currents plays a major role in determining the climate. Ocean circulation is responsible for increased surface temperature at high latitudes, reduced snow and sea-ice coverage, and reduced sensitivity to daily and yearly changes in atmospheric carbon dioxide concentrations. This might be why no strong link has yet been found between the rise in carbon dioxide levels and temperatures.

THE SOLAR INFLUENCE

From 1645 to 1715, there was apparently a minimum of sunspot activity (FIG. 4-5). This is known as the Maunder Minimum, the nineteenth century English astronomer Walter Maunder who discovered it in 1894. It was blamed for a span of unusually cold weather in Europe and North America during what has been called the Little Ice Age. The cooling period began around 1500 and lasted for about 350 years. The beginning of the Little Ice Age was associated with an increase in volcanic activity. Volcanic ash might have blocked out the Sun in the higher atmosphere and lowered temperatures. Increasing levels of industrialization and the associated contribution of atmospheric carbon dioxide might have brought the world out of the Little Ice Age by contributing to the greenhouse effect. Otherwise, this period could have been a prelude to the next major ice age.

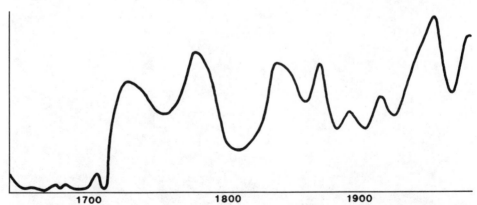

Fig. 4-5. Sunspot peaks through time. Note the apparent minimum activity ending in 1715.

Several correlations between the Sun and weather have been attempted. Perhaps the strongest link so far is the effect of sunspots on the vortex of stratospheric winds that swirl over the North Pole during the winter. The polar vortex breaks down when the wind in the lower stratosphere over the equator changes direction from west to east.

This change occurs only when there is a maximum number of sunspots. The more sunspots, the warmer the winter temperatures because of the breakdown of the polar vortex and the subsequent intrusion of warm air.

The sunspot cycle is a periodic occurrence of sunspots that ranges in duration from 9 to 14 years, with an average of 11 years. Other solar activity (FIG. 4-6) varies directly with the solar cycle, which has a period of about 22 years. The polarity of the Sun's magnetic field reverses every other sunspot cycle, or about every 22 years. Sunspots are associated with solar magnetic fields that are several thousand times stronger than the magnetic field on the Earth's surface.

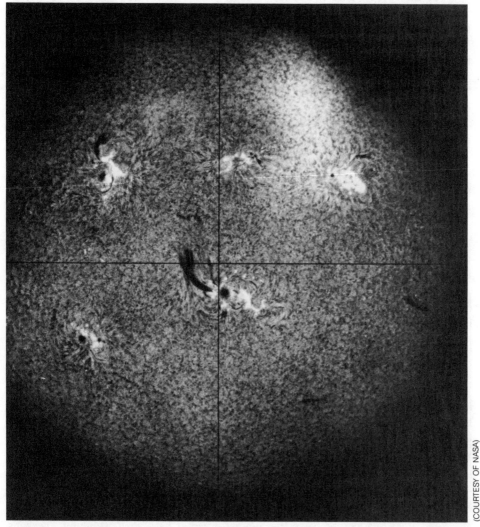

(COURTESY OF NASA)

Fig. 4-6. Sunspots and other solar activity viewed from the solar telescope on board Skylab.

The solar cycle might be the result of a global oscillation of the Sun that rhythmically changes its size by a minute amount, which results in slight variations in solar output. The oscillations also seem to affect the climate on Earth. Studies of drought in North America have shown a drought cycle of roughly 22 years, which seems to correspond to the solar cycle. Apparently, the climatic variation results when charged particles from the Sun modulate the Earth's ionosphere, which is the electrified portion of the upper atmosphere. The tropopause, the boundary between the troposphere and stratosphere, appears to rise and fall over the equator in rhythm with the solar cycle. The height of the tropopause is related to the strong convection beneath it. This indicates that there might be a direct connection between solar activity and the weather.

The *Solar Maximum Mission* satellite (FIG. 4-7), which began monitoring the Sun in 1980, measured a decrease in solar irradiance, or brightness, of nearly 0.1 percent between 1981 and 1984. If a decrease in luminosity as large as this lasted as long as a decade, it could have a major effect on the global climate. The apparent decrease in solar output might have been caused by a decrease in the number of sunspots since the beginning of the decade.

(COURTESY OF NASA)

Fig. 4-7. Artist's concept of the space shuttle crew repairing the Solar Maximum Mission satellite.

Sunspots do not block out the Sun, but instead are an indication of increased solar activity. Sunspots appear dark because their large diameters obstruct the convective flow of heat toward the surface of the Sun. Sunspots uncover the hotter depths and expose more solar radiation to space. Less obvious, but more extensive, are bright areas called plage, which accompany the sunspots. The sunspot cycle reached a minimum in 1975, a maximum in 1981, another minimum in the autumn of 1985, and another maximum in the winter of 1989. The Sun's irradiance starts to increase at the beginning of each new cycle, which warms the Earth by a small fraction of a degree.

The total amount of solar energy intercepted by the Earth, if spread uniformly across the surface and averaged over a year's time, is known as the average solar input. It is about 1,000 watts per square meter, or square yard, at any given moment. The so-called solar constant suggests that the amount of solar energy impinging on the Earth has remained fairly steady throughout time. This is not the case, however. The Sun's luminosity has been steadily increasing since the very beginning. It is only because the increase is so minute that scientific instruments have yet to detect it.

CLOUD FORMATION

Warm air has a greater capacity to hold water vapor than cold air. As moist, warm air ascends, it cools to the saturation point of water vapor, which can condense into a liquid even before the air reaches the saturation point. When the saturation point is reached, however, condensation will not necessarily take place if condensation nuclei are not present. The nuclei around which water vapor condenses are atmospheric dust particles composed of volcanic ash, wind-blown clay particles, salt from ocean spray, soot from forest fires, micrometeorites, and more recently, pollution from human activities. Without condensation nuclei, the air would become supersaturated with water vapor and no clouds would form.

Clouds are formed when a moisture laden parcel of air is heated and then rises through the atmosphere. As the air slowly ascends, the atmospheric pressure gradually decreases, which causes the air to expand. The energy needed for this expansion comes from within the parcel of air itself in the form of heat loss, which results in a drop in temperature. The rate of temperature loss is called the adiabatic lapse rate. It is about 1 degree Celsius per 100 meters or 5 degrees Fahrenheit per 1,000 feet of climb (FIG. 4-8).

A parcel of moist air that continues to rise eventually reaches the dew point. Water vapor then condenses around minute dust particles. As the water vapor condenses, it releases latent heat, which is heat given off when water changes state from a gas to a liquid or from a liquid to a solid. The release of latent heat in turn slows down the cooling of the air. Once the saturation point is reached, the air becomes more buoyant because moist air is more buoyant than dry air. This explains why clouds grow upward and are associated with the ascending air of low-pressure systems.

A comparison of the number of cloudless days—when an average of 10 percent or less of the daytime sky was obscured by clouds, fog, haze, or smoke—was made for 45 major cities in the United States between 1900 and 1982. The results indicate that the second half of this century has had more cloudy days than the first half. Several reasons might explain the trend toward more cloudiness. One suggestion is that carbon soot in

The 1988 forest fire at Yellowstone National Park, Wyoming.

Destruction to Yellowstone National Park from that fire, which burned nearly half the park.

Victoria Valley, showing valley glacier, Victoria Land, Antartica.
Melting of these glaciers causes flooding in other parts of the world . . .

. . . Coastal cities are vulnerable to the rising sea . . .
Eroded seawall at the base of the Cape Hatteras lighthouse, North Carolina.

. . . And delicate wetlands are reclaimed by the rising sea.
Swamp region of the Everglades National Park, Florida.

Star dunes in the Namib Desert, South West Africa.
At the present rate of destruction, forested lands will become deserts
by the middle of the 21st century.

Contaminated water at the Argo Tunnel at Idaho Springs, Colorado.

The mouth of the Big Thompson Canyon after the flood of July/August 1976.

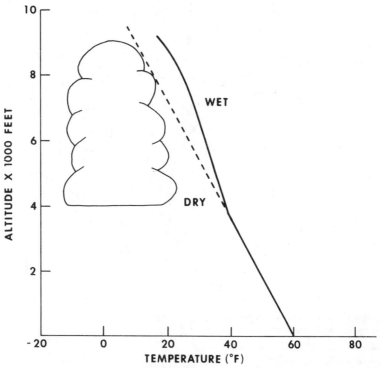

Fig. 4-8. The adiabatic lapse rate.

the contrails of jet aircraft acts as condensation nuclei and encourages cloud growth. Routes taken by ships at sea can be detected on satellite imagery because their exhaust smoke acts as condensation nuclei, which leaves behind a thin trail of clouds. Other sources of air pollution could also supply the microscopic particles for the condensation of cloud droplets.

Clouds play a most important role in setting the Earth's temperature. Air pollution could cause an increase in cloudiness, which might cause more sunlight to be reflected back to space so that the Earth would cool. At the same time, increased atmospheric carbon dioxide concentrations could cause a greenhouse effect, which would trap heat and warm the Earth. In the end, these two processes might cancel each other out to some degree.

PRECIPITATION

Once tiny cloud droplets form, several forces must come into play before precipitation falls. Ordinary raindrops are several million times larger than cloud droplets (FIG. 4-9). Most rain begins as ice. Water droplets can also be supercooled well below the freezing point without turning into ice because the microscopic water droplets are too small to freeze into a cohesive crystal structure. Ice also attracts water vapor because the vapor pressure near ice is lower than it is near water. This causes water molecules to leave the cloud droplets and flow toward the ice, which grows by the addition of more water molecules.

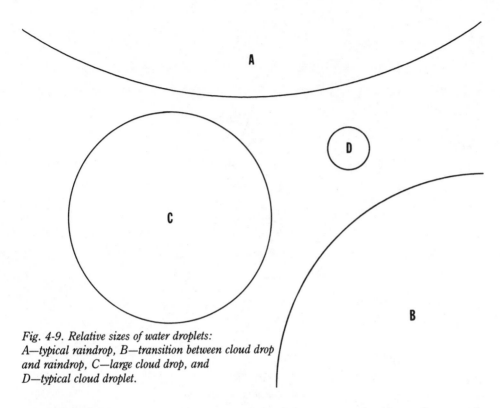

Fig. 4-9. Relative sizes of water droplets:
A—typical raindrop, B—transition between cloud drop
and raindrop, C—large cloud drop, and
D—typical cloud droplet.

Clouds that rise high into the thin, cold air of the upper troposphere often contain both ice crystals and supercooled water droplets. The ice crystals grow at the expense of the water droplets until they are heavy enough to fall through the cloud. When the ice crystals reach the warmer air below the cloud, the ice melts and becomes rain. In the winter or in the polar regions, the air nearer the ground might be too cold to melt the ice crystals, so the precipitate falls as sleet or snow (FIG. 4-10).

Clouds that originate in the tropics are rarely cold enough to permit the formation of ice crystals and must rely on some other mechanism for water droplets to merge into raindrops. Water droplets of uniform size rarely collide. When they do, they usually rebound due to surface tension, which produces a thin film over the droplets. In order for raindrops to form, some of the water droplets must have started out rather large. This is accomplished by the condensation of water vapor around large salt particles.

The salt particles are produced by salt spray from ocean waves and the bursting of air bubbles on the surface of the sea. These particles are sent aloft by tropical air currents and attract water vapor much like a salt shaker becomes moist and refuses to pour on a humid day. The salt particles produce larger, heavier cloud droplets. These are the first to overcome the lifting power of rising air currents. On the way down, they collide and combine with smaller droplets.

Not all precipitation reaches the ground. The same warm updrafts that created the clouds in the first place might evaporate the rain or snow soon after it leaves the cloud.

Fig. 4-10. A precipitation and air temperature gauge at Wolverine Glacier, Alaska.

This produces what is called virga. As long as there exists an upward movement of air, precipitation is inhibited. This is because the release of latent condensation heat keeps the rising air warmer than the surrounding air and maintains its upward movement. The air keeps rising until it enters a layer of warmer air, in which the rising parcel of air no longer has buoyancy.

As long as the parcel of air maintains its buoyancy, the upward movement of air within a cloud retards the fall of precipitation until the raindrops or snowflakes grow large and heavy enough to overcome the updraft. This produces a reoccurring precipitation-evaporation cycle within the cloud. Eventually, a few large drops do fall to the ground, cool the surface, and cause the updrafts to cease and a downpour to commence.

TROPICAL STORMS

Tropical cyclones are nature's most spectacular and most destructive storms. They begin as tropical depressions over the oceans between latitudes 5 and 20 degrees north and south. The sea surface temperature must be above 80 degrees Fahrenheit for strong convection to occur. The storms must also form some distance away from the equator in order for the Coriolis effect to provide the necessary spin. Therefore, tropical cyclones, or typhoons, usually occur in the summer and autumn when the Sun can heat the sea well to the north or south of the equator.

Only 1 out of 10 tropical depressions develop into a tropical cyclone, which typically has a life-span of 7 to 10 days. Once a cyclone forms, the storm is driven by the trade winds and normally heads west, which places the western shores of continents at the greatest risk (FIG. 4-11). The western Pacific Ocean has the highest frequency of tropical cyclones, which take more lives and cause more damage than in any other part of the world. Japan and the Philippines are particularly hard hit because these countries sprawl across the paths taken by many typhoons.

(COURTESY OF NOAA)

Fig. 4-11. Weather satellite view of Hurricane Diana off the coasts of Georgia and South Carolina.

The majority of tropical cyclones have a diameter of 300 to 400 miles. Some have been known to span 1,000 miles or more across. Cyclones rotate counterclockwise in the Northern Hemisphere and clockwise in the Southern Hemisphere. These patterns follow the Earth's rotation when viewed from above each corresponding pole. The circulating winds spiral in toward the center or eye at speeds of over 100 miles per hour. The eye ranges from 5 to 25 miles in diameter. The winds within the eye are fairly calm and the skies are nearly clear. The extreme low pressure in the eye sucks seawater into a mound several feet high. Winds of 100 miles per hour or more push the water ahead of the tropical cyclone, which produces a storm surge or tidal wave that can severely erode beaches and wreck shore property (FIG. 4-12).

(COURTESY OF NOAA)

Fig. 4-12. Hurricane storm surge damage to homes at Virginia Beach, Virginia, in March 1962.

When a tropical cyclone reaches land, it is deprived of its primary energy source, which is warm, moist air from the sea. Instead, it must generate energy from latent heat produced by heavy rainfall. Tropical cyclones are an important source of rainfall throughout many parts of the world. Once a cyclone reaches land, between 3 and 6 inches of rain is common and upwards of several tens of inches is possible within a 24-hour period. This can result in severe flooding (FIG. 4-13).

After a tropical cyclone has traveled some distance over land, it loses its energy and turns into an ordinary low-pressure system. In the Atlantic Ocean, these low-pressure centers, or cyclones, are an important source of rainfall for Europe. Those that spawn in the Pacific Ocean bring much needed rains to North America.

Fig. 4-13. Flooding in New Orleans, Louisiana, on October 30, 1985 due to Hurricane Juan.

THE MONSOONS

Nearly half the world's population depends on the monsoon rains for survival. Monsoons are seasonal changes in wind direction that alternately produce wet summers and dry winters. During the rainy season, there are periods of drenching squalls interspersed with equal periods of a week or so of sunny weather. During the monsoon's dormant phase, the weather is hot, dry, and stable with an absence of tropical storms.

Monsoons owe their existence to the difference in temperature between ocean and land. This in turn produces a difference in atmospheric pressure that must be equalized by winds. Water evaporating from the oceans at any given time stores about one-sixth of the solar energy reaching the surface of the Earth. During a monsoon, part of this enormous reservoir of solar energy that has collected over the oceans is released over the land when the water in moist ocean air condenses into rain.

The specific heat, or the amount of stored heat, of water is more than twice that of land. The temperature of dry land will therefore increase more than twice as rapidly as the same mass of ocean when exposed to an equal amount of solar radiation. Since the Earth's surface is three-quarters water, the oceans absorb a great deal of heat. The ocean's greater heat capacity relies on its efficiency in mixing heat at great depth. Soil and rocks, on the other hand, can only store heat to a depth of a few feet. The ocean distributes huge quantities of heat throughout a large mass of water. Because of its high specific heat and great mixing ability, the temperature of the ocean surface varies much less than that of land. The oceanic variance may be no more than about 25 degrees Fahrenheit between seasons.

The summer monsoons continue for as long as there are unbalanced forces between land and ocean. As fall arrives, the temperature of the ocean drops and diminishes the temperature difference between land and sea. The energy of the system then runs down, the monsoon retreats, and the winter dry season begins. With the onset of winter, the land loses heat much faster than the ocean. The resulting increase in heat loss from the land and the greater heat capacity of the ocean causes the winds now to blow in the opposite direction.

If the monsoons should fail due to a climatic disturbance brought on by greenhouse warming, a major drought could occur which would be a horrendous tragedy.

5

Fossil Fuel Combustion

L ESS than a quarter of the human population consumes some 80 percent of the world's resources. A few rich nations are more responsible for the pollution and the degradation of the environment than all the rest of the world combined. Consumption of resources and the production of pollution per capita in the U.S. is roughly four times greater than that of other industrial nations. It is no wonder that developing nations are rushing to employ their own resources so they can participate in the prosperity.

Unfortunately, the additional burden of atmospheric pollution and carbon dioxide from developing nations might worsen the problem of greenhouse warming. The increased efficiency of energy use and alternative fuels, however, might help developing countries raise their standard of living without a significant increase in energy use or a corresponding increase in pollution.

OIL CONSUMPTION

The exploitation of coal, steam, and iron produced the Industrial Revolution. In the United States, the use of coal gave way to oil and natural gas after the discovery of oil in Titusville, Pennsylvania, in 1859. The search for oil even rivaled the California gold rush that took place a decade earlier. The landscape became dotted with oil derricks. Vast fortunes were made, which produced a new generation of wealth.

Interest in offshore drilling for oil and natural gas in shallow coastal waters began in the mid-1960s. Drilling stepped up considerably a decade later after the 1973 Arab oil embargo when American motorists were forced to wait in long lines at gas stations.

Over the last two decades, offshore drilling has become extremely profitable. Important finds such as Prudhoe Bay on Alaska's North Slope and on the North Sea off Great Britain were the result of intensive exploration efforts. The desire for energy independence also sent oil companies exploring for oil in the deep oceans where many difficulties were encountered. Storms at sea and the loss of personnel and equipment could not justify the few discoveries that were made, however.

About 20 percent of the world's oil and about 5 percent of its natural gas production is offshore. Projections indicate that perhaps twice as much oil will be pumped from the seas than from the land in the future. Unfortunately, some of the offshore oil leaks into the oceans. As much as 2 million tons spill each year. This could create enormous environmental problems as production increases.

In the early 1980s, the Department of the Interior estimated that large reserves of oil and natural gas remain to be discovered in offshore deposits large enough to be commercially exploited around the United States. However, by the mid-1980s, the Department cut in half its estimate of oil reserves in offshore fields. The revised figures reflect the fact that oil companies came up practically empty-handed after four years of exploratory drilling in highly promising areas of the Atlantic Ocean and off the coast of Alaska. It was hoped that these regions would help keep the nation well supplied with oil because without it the United States could become dangerously dependent on foreign sources.

Reservoirs of oil and natural gas require special geological conditions. There must be a sedimentary source for the oil, a porous rock to serve as a reservoir, and a confining structure to trap the oil. The source material is organic carbon in fine-grained, carbon-rich sediments. Porous and permeable sedimentary rock such as sandstones and limestones serve as a reservoir. Geological structures that result from folding or faulting of sedimentary layers can trap or pool the oil (FIG. 5-1). Oil is also often associated with thick beds of salt. Because salt is lighter than the overlying sediments, it rises toward the surface and creates salt domes that help trap the oil.

Most of the organic material that forms oil comes from microscopic organisms that primarily originate in the surface waters of the ocean and are then concentrated in fine particulate matter on the ocean floor. In order for organic material to be transformed into oil, the rate of accumulation must be high or the oxygen in the oceanic bottom water must be low because the material cannot be oxidized before it is buried under thick layers of sediment. Oxidation causes decay, which destroys organic material. Areas with a high rate of sediment accumulation that are also rich in organic material are the most favorable sites for the formation of oil-bearing rock.

After deep burial in a sedimentary basin, the organic material is heated under high temperatures and pressures, which chemically alter it. In essence, the organic material is converted into hydrocarbons by the heat generated in the Earth's interior (FIG. 5-2). If the hydrocarbons are overdone, gas results.

Hydrocarbon volatiles and seawater that are locked up in the sediments migrate upward through permeable rock layers and accumulate in traps formed by sedimentary structures, which provide a barrier to further migration. In the absence of such a cap rock, the volatiles continue rising to the surface and escape. It takes from several tens of millions to a few hundred million years to produce oil. Production mainly depends on the temperature and pressure conditions within the sedimentary basin. Plate tectonics plays

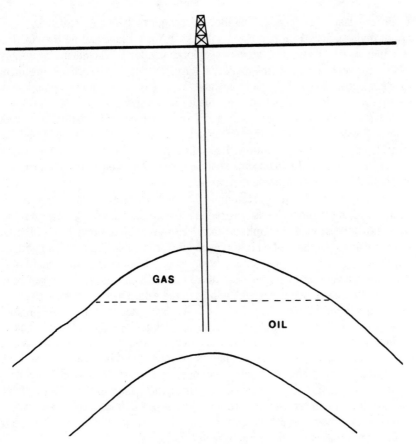

Fig. 5-1. A typical oil trap.

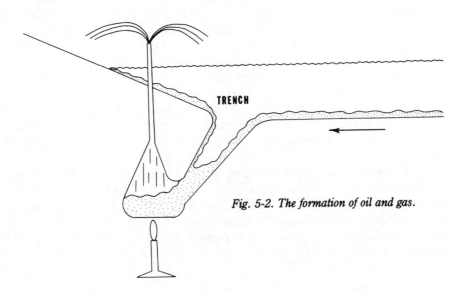

Fig. 5-2. The formation of oil and gas.

Fig. 5-3. Derrick hands drilling an oil well. Only one out of ten wells drilled actually produce oil.

an essential role in determining whether these conditions are met and greatly aids oil companies in their exploration activities (FIG. 5-3).

Soon after the turn of the century, the oil supply will begin to fail to meet the growing demand placed upon it by an increasing number of industrialized nations. By the year 2035, it is estimated that most, if not all, of the world's petroleum reserves will be depleted. Of the over one-trillion barrels of oil thus far discovered, one-third or more has already been consumed. As oil production levels off and then falls, alternative fuels will have to be developed to meet the demand for energy, which will continue to grow even with conservation methods in place. For oil importing countries, this will require a transition from a dependence on oil to a greater reliance on other fossil fuels, nuclear energy, and renewable energy sources.

OIL SPILLS

Offshore oil spills are perhaps the most hideous and damaging of all coastal pollution. In January 1969, a major oil well blowout at Santa Barbara, California, released a tremendous quantity of crude oil into the Pacific Ocean. The local ecological consequences were immediate and disastrous. This dramatic event made the public aware of the enormously high level of industrial pollution in both the atmosphere and the waters. It also encouraged Congress to enact legislation that limits industrial pollution of the nation's waters.

Major oil spills, however, continue to occur. The 1976 *Argo Merchant* spill off Cape Cod (FIG. 5-4) led to a reexamination of oil spill contingency plans designed to protect productive fishing grounds. Another major oil well blowout in the Gulf of Mexico in June 1979 fouled Texas beaches. A massive oil spill in the Persian Gulf, which resulted from the long Iran-Iraq war, has threatened to endanger the entire ecology of the Persian Gulf area and could have serious consequences for decades to come.

Fig. 5-4. The Argo Merchant *oil spill off Cape Cod in December 1976.*

The March 24, 1989 grounding of the oil tanker *Valdez* in Alaska's Prince William Sound was the nation's worst oil spill. Some 11 million gallons of crude oil leaked from the torn side of the ship. Only 10 percent of the lost oil was recovered. The rest created an oil slick that covered a large portion of the sound. The spill will have a lasting affect on the local ecology and economy. Cleanup efforts were hampered by the lack of equipment and bad weather.

In most cases, an oil slick is contained with floating booms or by labor intensive cleanup (FIG. 5-5). Detergents or other chemicals can be used to break up the oil, but these chemicals might cause additional damage to the ecology. Oil slicks are transported by ocean currents, local winds, and tides. The success of the cleanup operation ultimately depends on accurate weather forecasts, particularly wind speed and direction. There is a critical need for research into the dispersion, spreading, and subsurface transport of oil slicks. This information would help cleanup crews determine the extent of the oil spill and the fate of the coastal ecology.

Satellites are able to detect and monitor oil slicks because the oil has a higher index of refraction than the surrounding water at ultraviolet and visible wave lengths. In the thermal infrared region, oil has different radiation characteristics than water. Oil slicks dampen small waves and thereby reduce the amount of radar backscatter, which can be detected on satellite radar imagery. With continual surveillance, satellites can aid cleanup efforts by tracking the drift and dispersion of oil slicks.

Fig. 5-5. Workers clean the beaches of Sandy Hook, New Jersey, from a major oil spill in March 1980.

DIRTY COAL

Since it is unlikely that the consumption of fossil fuels will decline in industrial countries over the next 30 years, other safe alternatives such as fission, fusion, and geothermal energy need to be developed, or industrial plants will have to convert to coal, a particularly dirty fuel.

Efforts have been made to clean up coal-fired plants by installing scrubbers that eliminate sulfur dioxide emissions and using low-sulfur coal. However, many older power plants and power plants in other countries are not required to make these expensive investments. Thus, sulfurous smoke continues to cloud the skies and produce acid precipitation.

Prior to the industrial era, the small fires of civilization had virtually no effect on the atmosphere. Populations were small and the fuel burned was mostly wood. As time progressed and populations grew, however, the forests of Europe rapidly disappeared until the timely discovery of coal. Man probably first stumbled upon coal when exposed seams were set ablaze by lightning. It is known that at an early time, native Americans used coal for their cooking fires.

Although the Industrial Revolution began in Britain and Europe in the mid-1700s, it did not take hold in the United States for almost another century because of this country's mostly agrarian economy. During the Industrial Age, there was a proliferation of inventions, including steam locomotives and steamships. These vehicles transported the materials needed to feed the voracious appetite of heavy industry. In every major city, smoke and soot belched out of numerous smokestacks. Although people generally benefited economically from this prosperity, they also suffered serious side effects.

Smoke was so heavy in some major cities that people died from lung ailments in unusually large numbers. Ash and soot covered buildings, trees, and other objects, which were also blackened by the smoke. Smoke stung the eyes, and the sulfurous stench of burning coal irritated the nose. Some days the smoke was so heavy that the Sun was barely visible. Nonetheless, many farmers quit their fields and fled to the cities to find factory jobs. By the mid-nineteenth century, the populations of the major industrialized nations were becoming more urban than rural for the first time in civilization.

By the beginning of the twentieth century, industrial plants began converting to oil and natural gas because these were more convenient fuels than coal. Not only were these fuels easier to utilize and more efficient, they were also less polluting. Oil and natural gas produce only about half the amount of carbon per unit of energy as coal (FIG. 5-6). Electricity generated by fossil fuels also became a new source of energy, which revolutionized industry and transportation. However, due to the higher cost and future scarcity of petroleum, the trend appears to be reversing and more coal is now being used for the generation of electricity. This reversal brings with it all the problems encountered in the past.

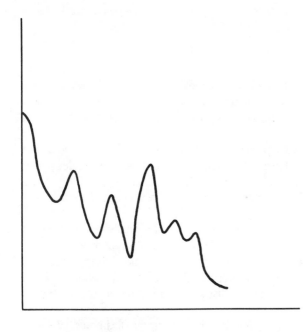

Fig. 5-6. Relative smoke levels have been dropping during most of this century due mostly to the conversion from coal to petroleum.

Since the decade of the 1970s, the United States has increased its consumption of coal by about 70 percent, mostly for the generation of electricity in huge coal-fired plants. The generation of electricity accounts for about three-quarters of the coal consumed in the U.S.

Because coal from eastern underground mines has a high sulfur content, most plants use western coal, which is mined from massive open pits (FIG. 5-7). In order to keep up with demand, the U.S. will have to mine about 60 percent more coal by the turn of the century. Presently, the total world coal production is approximately 5 billion tons annually. The U.S. accounts for about half of the coal mined and consumed by the free world.

(PHOTO BY P.F. NARTEN, COURTESY OF USGS)

Fig. 5-7. Open-pit coal mining at the West Decker mine in Montana.

The world's coal reserves far exceed those of all other fossil fuels and are sufficient to support large increases in consumption well into the next century. The amount of economically recoverable coal reserves is upwards of 1 trillion tons. At the present rate of consumption, these reserves could last well over 200 years. The U.S. possesses half of the free world's coal reserves (FIG. 5-8), which remain practically untouched. China and the Soviet Union have most of the rest. Since coal is the cheapest and most abundant fuel, it will become a favorable alternative source of energy. Unfortunately, coal-burning yields significantly more carbon dioxide per unit of heat than oil and natural gas.

It took millions of years to make the world's fossil fuel reserves. If consumption continues at its present furious pace, all the reserves might go up in smoke in a matter of a few decades. This could cause the atmosphere to overload on carbon dioxide. Millions of tons of sulfur dioxide and nitrogen oxides would also be discharged into the atmosphere each year. These gasses combine with oxygen and moisture in the atmosphere to produce acid rain, or they rise to the upper atmosphere and destroy the ozone layer.

Fig. 5-8. Location of coal deposits in the U.S.

ALTERNATIVE ENERGY

If alternative sources of energy are not found before fossil fuels begin to run out, industrial nations will face a crisis in the upcoming years. Atomic energy was once thought to be the answer to the world's energy problems because nuclear plants do not pollute the atmosphere and nuclear wastes can be managed properly. Strict government regulations and frequent construction delays, however, have caused the cost of these plants to skyrocket. Unless regulations can be streamlined and power plants constructed more efficiently, it is doubtful that nuclear energy will ever be an economical alternative in the U.S. In order to combat the greenhouse effect, however, a reassessment of nuclear energy might be required. Many countries, particularly France, rely heavily on nuclear energy to replace costly fossil fuels.

In April 1986, a tragic accident at the Chernobyl nuclear power station near Kiev in the Soviet Union caused an explosion and partial meltdown of one of the three uranium-graphite cores. This released a dangerous radioactive cloud that killed dozens of people and injured thousands more in western Russia and parts of Europe. It contaminated milk and other food stocks, including herds of reindeer, which are a major source of food for the Lapps who roam the arctic regions of northern Europe. The nuclear accident set world opinion against the proliferation of nuclear power plants and spurred the search for alternate energy sources.

One alternate source is solar energy. The Sun itself is a gigantic nuclear furnace that radiates a tremendous amount of solar energy. The sunlight that strikes the Earth's surface produces enough energy in one square meter to operate a 40-watt light bulb. Unfortunately, converting this energy into a usable form, such as electricity, is often difficult and inefficient.

One method of directly converting sunlight into electricity is the use of solar cells, which have an optimum efficiency of about 20 percent. The manufacture of solar cells, which requires nearly pure materials, is also very expensive. This makes their use uneconomical on a large scale. Less efficient solar cells might be manufactured in mass quantities to bring the price down. Promoters claim that home solar electrical plants can offer a long-term savings over the cost of running a power line to a rural house located a long distance from a utility pole. For the present, however, solar cells are largely used to power satellites and ground communication systems.

Another way to convert sunlight into electricity is to focus the sunlight into a powerful narrow beam by using banks of heliostatic mirrors, which automatically track the Sun as it moves across the sky. The light beam is focused on a central receiving station where the intensified light heats a boiler. The superheated steam then drives a steam turbine generator.

The most obvious drawback to solar generating stations is that they do not operate efficiently on a cloudy day and not at all at night. Therefore, solar power plants must be located in areas where the Sun shines for long periods throughout most of the year. Deserts are an ideal location because the land is also relatively inexpensive. Solar power stations still cannot compete economically with conventional fossil fuel generating stations. In the future, however, they might have their day in the Sun when fossil fuels become scarce.

Homes and apartment buildings can utilize solar energy to supplement conventional water heaters and furnaces (FIG. 5-9). This can reduce utility bills and conserve unrenewable energy resources. Areas such as the Southwest, which receive abundant sunlight, can take full advantage of solar energy. The systems usually pay for themselves in utilities savings in about 10 years.

Wind and water power are also forms of solar energy. The Sun drives the air and water currents that are responsible in large part for our weather. In windy locations, such as the coasts where the offshore and onshore wind currents are fairly reliable, huge electrical generating wind turbines have been built to harness this free energy (FIG. 5-10). The wind also drives the oceans' waves, which can be harnessed to produce electricity. Perhaps the most successful use of solar energy is hydroelectric power. Hydroelectric projects are extremely expensive, however, and the most accessible sites have already been utilized. Because of this, the use of hydroelectric power will probably not increase significantly in the future.

Geothermal is a promising source of energy that has been around for a long time. Like solar energy, the potential for geothermal energy is enormous. Unfortunately, the amount of this energy released to the surface is less than one-tenth of 1 percent of the energy received from the Sun. In a sense, the Earth's interior can be thought of as a natural nuclear power reactor because the heat is mainly derived by the decay of radioactive elements. Many steam and geyser areas around the world are generally associated with active volcanism and are potential sites for tapping geothermal energy for steam heat and electrical power generation (FIG. 5-11).

Fig. 5-9. Solar panels on houses help save on utilities.

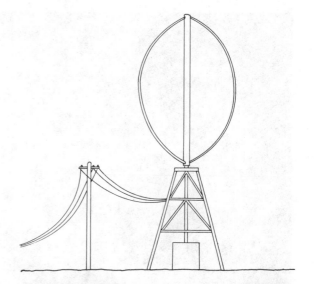

Fig. 5-10. A Darrieus-type wind generator.

Fig. 5-11. Geothermal generating plant at the Geysers in California.

(COURTESY OF USGS)

In a world of rapidly diminishing fossil fuels, an energy source that is both renewable and essentially nonpolluting is worth pursuing even at an enormous cost. Such an energy source is fusion (FIG. 5-12). The fuel used for fusion is deuterium or heavy hydrogen, which is abundantly available in seawater. The energy from the fusion of deuterium in a pool of seawater 100 feet square and 7 feet deep could provide the electrical needs of a city of a quarter of a million people for an entire year.

Fusion is safe. Its byproducts are energy and helium, a harmless gas that escapes into space. Research into the development of controlled thermonuclear fusion, however, has been plagued with failure and frustration, although many milestones have been passed, including the possibility of producing cold fusion. The enormous engineering and technical problems that still remain, have caused some experts to question whether energy from thermonuclear fusion will ever be economically viable. At the earliest, it is not expected to be productive until well into the next century.

(COURTESY OF U.S. DEPARTMENT OF ENERGY)

Fig. 5-12. A Tokamak fusion test reactor at Princeton University, New Jersey.

Man has been blessed with a world rich in natural resources. The exploitation of mineral and energy resources has improved the lives of most people. An unfortunate byproduct of man's improvements, however, is pollution and the health hazards it entails. The depletion of natural resources could also threaten future advancement. Through the conservation of natural resources, the wealth of the Earth can be preserved for future generations.

6

Industrial Pollution

BY using satellites to monitor the Earth, scientists are learning more about the state of the planet and of its intertwining global processes. With this knowledge, they have become aware of changes that are taking place and of the fact that man is responsible for many of these changes. The economic developments of large portions of the Earth has dramatically changed the patterns of land and water use. There has been large-scale extraction and combustion of fossil fuels and widespread use of man-made chemicals in industry and agriculture. These activities appear to be altering the cycles of essential nutrients in the biosphere. They also appear to be affecting the climate by altering precipitation patterns around the world.

Man currently consumes some 40 percent of the Earth's net primary production, which is the energy stored by green plants throughout the world. A doubling of the human population by the middle of the next century will require the consumption of twice the world's net primary production. This could be disastrous considering the destructive impact of the current level of human activities. Human populations are growing so explosively and are modifying the environment so extensively that we are inflicting a global impact of unprecedented dimensions. Man's activities could possibly cause greater havoc than the worst calamities the planet has ever endured.

AIR POLLUTION

Air pollution has become a growing threat to health because of the ever-increasing emission of air contaminants. Each day, an average adult inhales about 30 pounds of air.

In a lifetime, this could amount to enough air to fill an enclosed football stadium. If one can imagine that facility filled with polluted air, all of which is circulated through one individual's lungs, it would be easy to comprehend the necessity of having clean air.

Foreign chemicals and particulates injected into the atmosphere, either by natural or man-made sources, are air pollution. Natural pollutants include salt particles from breaking waves, pollen and spores released by plants, smoke from forest fires, wind-blown dust, meteoritic dust, and volcanic ash (FIG. 6-1). Humans are by far the greatest polluters. Since the industrial era began, people have rivaled nature as the depositors of the greatest amount of toxic wastes and particulates into the atmosphere.

Fig. 6-1a. The May 18, 1980 eruption of Mount St. Helens, Washington.

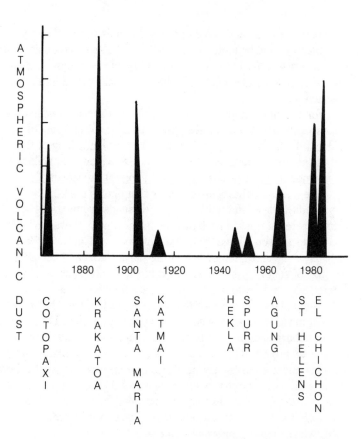

Fig. 6-1b. The relative load of volcanic ash injected into the atmosphere over the last hundred years.

The industrial age brought prosperity and a rapid growth of the world's population, which in turn required additional industrialization. With more mouths to feed, more forests were felled and burned to make room for additional agriculture. This rapid-paced industrialization might have provided greater economic security, but it also produced increased amounts of pollution and a consequential reduction in public well being.

Air pollution is particularly hazardous if it is not diluted by the atmosphere. Unstable air carries smoke and exhaust fumes upward on air currents where they are mixed with cleaner air aloft and disbursed by the winds. Because of the direct effect of wind speed, the concentration of pollutants is half as much with a wind of 10 miles per hour than with one of only 5 miles per hour. High pollution days do not necessarily mean that there is an increase in pollution output, but that the air into which the pollution is released is not disbursed by the wind. This is what happens during a temperature inversion. Warm air overlies cooler air and acts like a lid that prevents the upward movement of air currents and traps the pollutants near the ground.

Atmospheric pollutants are grouped into primary pollutants, which are those emitted directly from primary sources such as smokestacks and motor vehicle exhaust pipes, and secondary pollutants, which are produced by chemical reactions that take place among the primary pollutants. Many reactions that produce secondary pollutants are triggered by sunlight and are called photochemical reactions. Other secondary pollutants are corrosive acids and deadly poisons such as ozone. Nitrogen oxides produced in factory furnaces and by motor vehicles absorb solar radiation and initiate a chain of complex chemical reactions. In the presence of organic compounds, these reactions result in the formation of a number of undesirable secondary products that are very unstable, irritating, and toxic.

For the past couple of decades, the release of cancer-causing chemicals into the atmosphere has been on the rise. Thousands of tons of dangerous chemicals are released by factories around the world each year. Many of these substances are rained out of the atmosphere and end up in rivers, lakes, and soils where they concentrate in toxic amounts through chemical and biological processes.

The amount of particulate matter that enters the atmosphere also has been increasing steadily. Slash and burn agriculture sends tremendous amounts of smoke into the atmosphere as does dust blown up from newly plowed or abandoned fields. Factory smokestacks send huge quantities of soot and aerosols into the air. Motor vehicle exhaust alone accounts for half of the particulates and aerosols in the atmosphere. The amount of soot and dust suspended in the atmosphere at any one time as the result of human activity is estimated to be about 15 million tons.

Much of the air pollution that reduces visibility and harms plants and animals as well as man-made structures is composed of dry deposits. These atmospheric particles consist of unburned carbon, dust particles, and minute sulfate particles. The finest particles are mainly produced by chemical processes that take place during the combustion of fossil fuels and are essentially acidic. High-temperature combustion yields nitrogen oxide along with gaseous nitric acid. Because of their small size these particles, called aerosols, scatter light that normally would heat the ground. Instead, the sunlight heats the atmosphere and causes a temperature imbalance between the atmosphere and the surface, which creates abnormal weather patterns.

The coarse particles in the atmosphere are derived mainly from the mechanical breakup of naturally occurring substances such as those produced by volcanic eruptions and dust storms. Large particles of carbon, called soot, are produced by forest fires (FIG. 6-2) and inefficient fuel combustion in wood-burning stoves. Under an atmospheric inversion, smoke from fireplaces can produce a persistent haze. Even in pristine areas, such as the arctic tundra where the air is considered clean, significant levels of sulfate-containing particles have been found. Studies show that these pollutants came from distant industrial sources.

In an effort to eliminate local air pollution problems from coal-fired plants, tall smokestacks were built so that pollutants would readily mix with the turbulent air above and be carried aloft. Unfortunately, this action only turned a local pollution problem into a regional one by allowing pollutants to travel for long distances.

All air pollutants clog the skies and reduce the amount of sunlight that reaches the ground, which subsequently cools the Earth's surface. At the same time, sunlight strikes airborne particulates and heats the atmosphere, which causes a thermal imbalance. One of the reasons that the increased level of carbon dioxide in the atmosphere has yet to result in a large upward trend in world temperatures is that it is offset slightly by the cooling effects of particulate matter in the atmosphere.

Fig. 6-2. A forest fire in Washoe County, Nevada, that destroyed approximately 1,500 acres.

WATER POLLUTION

The cycle of water flow from ocean to land and back to sea again, known as the hydrological cycle, cleanses the Earth of its natural and man-made pollutants by the process of dilution (FIG. 6-3). This process works only to a certain extent. Toxic wastes can become concentrated at lethal levels. Many toxic substances that are diluted to supposedly safe levels in streams, lakes, and the ocean are concentrated by biological activity.

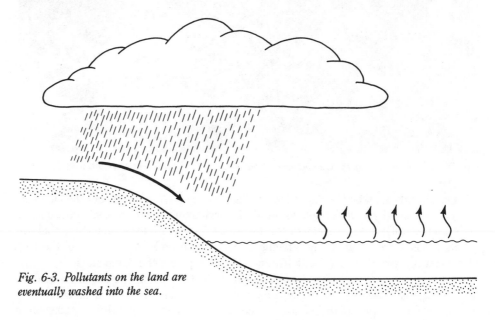

Fig. 6-3. Pollutants on the land are eventually washed into the sea.

The accumulation of toxic substances starts at the bottom of the food chain and works its way up to fish and other aquatic life, some of which are sources of food for humans.

Approximately 8 million tons of toxic waste is dumped into rivers and coastal waters each year (FIG. 6-4). Some of these toxic pollutants are powerful carcinogens and mutagens. Many are nonbiodegradable and persist in the environment for long periods. Hospital waste that could cause disease is dumped into the sea as is untreated or partially treated sewage.

(PHOTO BY W.R. BRAM, COURTESY OF USDA-SOIL CONSERVATION SERVICE)

Fig. 6-4. Water pollution in the Cumberland River north of Nashville, Tennessee, on May 12, 1970.

Ocean currents tend to bring the waste back to shore. As a result, beaches in many parts of the world are unsightly and unsafe for swimming. Waste is also concentrated between thermal layers and ocean fronts, the areas in which some of the most productive fishing grounds are found. In the Atlantic Ocean, the meandering currents of the Gulf Stream, which are often laden with fish, actually sweep over the dump sites.

Oil spills are by far the most damaging of all coastal pollution. The increased demand for offshore oil, collisions and groundings of oil tankers, and attacks on oil tankers by warring nations have led to disastrous ecological consequences. Heavy spills often require extensive cleanup efforts, especially in productive fishing grounds. In addition, marine pollution does not remain localized in highly contaminated areas such as the Mediterranean and the North Sea, but is spread to other parts of the world by ocean currents.

One of the most important sources of water for industrial, agricultural, and public purposes is groundwater (FIG. 6-5). Throughout the U.S., most of the water comes from aquifers that spread out underground for millions of square miles. Pollutants from landfills, storage tanks, industrial waste, agricultural chemicals, and low-level nuclear waste are percolating down through layers of sediment into the nation's aquifers. Some of these contaminants are carcinogenic. Even low levels of these substances can cause cancer. What were once nearly pure sources of water are becoming increasingly polluted. Many wells are now so polluted that they must be abandoned or, if possible, treated at the well head.

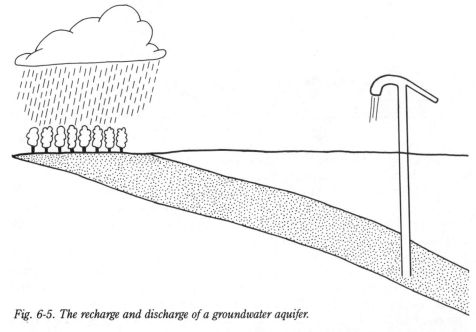

Fig. 6-5. The recharge and discharge of a groundwater aquifer.

Presently, up to 10 percent of the nation's groundwater supply is already contaminated. In the coming years, a quarter or more of the groundwater might be unusable. Because the pollutants flow so slowly, several years may pass before they appear in water wells. The groundwater problem has become the primary environmental challenge of the century. Cleaning up the contamination will be expensive, difficult, and in some cases impossible.

ACID RAIN

Industrialization brought with it the combustion of high-sulfur coal and oil and the smelting of sulfide ores, particularly in the heavily industrialized and urbanized temperate regions of the Northern Hemisphere. Human activity accounts for roughly 10 times more sulfur being injected into the atmosphere than natural sulfur emissions from sources such as volcanoes. The combustion of sulfur produces sulfur dioxide, which enters the atmosphere, combines with oxygen, and yields sulfur trioxide. This in turn combines with atmospheric moisture to produce sulfuric acid, which precipitates as highly corrosive acid rain (FIG. 6-6).

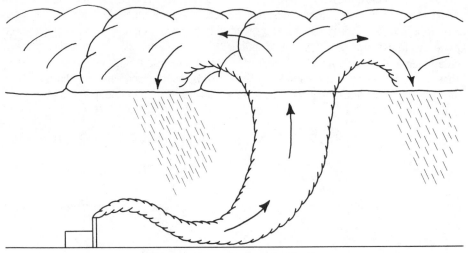

Fig. 6-6. The production of acid rain.

Measurements of the acidity of rain and snow reveal that in parts of eastern North America and northwestern Europe precipitation has changed from a nearly neutral solution at the beginning of the industrial era, two centuries ago, to a diluted solution of sulfuric and nitric acid. In the most extreme cases, rain has had the acidity of vinegar. Current research efforts in North America and Europe are directed toward determining both the direct and indirect effects of increased amounts of acids in the environment.

Acid rain has been known to exist for many decades in the vicinity of large cities and industrial plants. Tall smokestacks designed to disburse emissions high into the atmosphere and away from the cities also help the pollutants travel long distances, even across international borders. Coal-fired electrical generating plants in the Ohio valley, for example, produce acid rain clouds that travel into eastern Canada where the acid rain is destroying forests and once productive fishing streams and lakes.

Streams and lakes in other parts of the world, especially those that cover granitic strata and are not buffered by carbonate rocks, have become so acidic or polluted that fish populations have been decimated (FIG. 6-7). In the Adirondack Mountains of New York state, 90 percent of the lakes with high acid levels are completely devoid of fish. In Sweden, it is estimated that more than 15,000 lakes no longer contain fish populations due to acid rain.

Acid rain is especially harmful to aquatic organisms because it lowers the water's pH value. In seawater, the damage comes from nitrogen oxides. Nitrogen acts as a nutrient that promotes the growth of algae, which blocks out sunlight and depletes the water of dissolved oxygen. As a result, aquatic plants and animals suffocate. There has been a

Fig. 6-7. Water pollution is an unfortunate consequence of industrialization.

(COURTESY OF NOAA)

widespread increase in nitrate levels in the oceans along with higher concentrations of toxic metals, including arsenic, cadmium, and selenium. The main factors contributing to this increase are fertilizer and pesticide runoff and acid rain, which dissolves heavy metals in the soil.

Some soils have become so acidic they can no longer be cultivated. Plants are damaged by the adverse effects of acid on foliage and root systems. Acid precipitation is destroying the great forests of North America, Europe, China, and Brazil. There are widespread reductions in the width of tree-rings and an increased mortality rate for red spruce trees in the eastern United States. Resorts and wilderness areas, like those in the western United States, Norway, and West Germany's famous Black Forest, are losing much of their natural beauty because of acid rain.

Acid rain is possibly the most well studied and least acted upon of any pollution problem. The source of the problem has been known for quite some time. The basic chemistry that turns industrial and motor vehicle emissions into acid rain is well understood. Governments, however, are slow to require expensive mandatory emission controls because companies might lose their competitive edge due to the costly investments.

WASTE DISPOSAL

We are running out of room for our garbage. This problem was exemplified when a garbage barge left Long Island Sound in the summer of 1987 and ended up traveling 6,000 miles to six states and three countries before returning to New York because no one wanted its cargo. Man-made organic chemicals, heavy metals, pesticides, and other toxic substances are constantly seeping into the ground from landfills, buried gasoline tanks, septic systems, radioactive waste sites, farms, mines, and a host of other sources. The sources are so diverse that often it is difficult to determine the major cause of the groundwater pollution. Routine monitoring of industrial waste lagoons or landfills (FIG. 6-8) reveals that the chemicals are not being contained, but are ending up in nearby water wells.

Fig. 6-8. Disposal and monitoring of hazardous waste materials.

The cost of disposing of toxic waste on land is escalating. Many coastal metropolitan areas have been forced to dump municipal and industrial wastes directly into the sea. Typically, along the East Coast barges of waste are taken out about 100 miles and dumped beyond the continental shelf. After a day or so, the pollutants are diluted to supposedly safe levels. The pollutants, however, tend to concentrate in regions where the seawater density changes, such as ocean thermoclines and fronts. These are the same areas where fish feed.

Incineration at sea is one of the few methods available that detoxifies certain hazardous wastes, especially those that are highly chlorinated. One plan is to use incinerator ships to burn toxic wastes off shore. It would only be is a stopgap measure for treating toxic liquids until better methods to reduce or recycle the wastes can be developed. Ocean incineration is only suitable for treating a small percentage of hazardous wastes. The chemicals destroyed in this manner, however, are among the most toxic. There still remains many unresolved questions concerning the potential risks to health and the environment.

The disposal of radioactive wastes has received much attention in recent years because of increasing concerns over long-term environmental effects and the proliferation of nuclear technology throughout the world (FIG. 6-9). High-level radioactive waste is among the most difficult to dispose of due to its intense nuclear radiation, heat output, and longevity. It is generally thought that the best place to store nuclear waste is underground (FIG. 6-10). Mine repositories, however, are very expensive and must be located in geological formations that will remain stable for a million years or more. The disposal site must also be guarded against intrusion and theft for countless generations.

Fig. 6-9. The Oconee nuclear power plant in Seneca, South Carolina.

Fig. 6-10. Underground nuclear waste disposal at the Nevada Test Site.

OZONE DEPLETION

Satellites that monitor the ozone concentration in the upper atmosphere (FIG. 6-11) have discovered that the ozone layer is being depleted. Every September and October since the late 1970s, a giant hole, about the size of the continental United States, opens up in the ozone layer over Antarctica. Long-term records show that ozone levels in the far northern latitudes have dropped roughly 5 percent over the last 17 years. The ozone depletion is believed to have a chemical origin. It is also believed that the chemicals are man-made.

The amount of chlorine monoxide, which destroys ozone molecules, was found to be 100 times the normal level in the Antarctic stratosphere. Polar stratospheric clouds composed of frozen water and nitric acid crystals help chlorine destroy ozone by fostering certain chemical reactions. For this reason, there is no comparable Arctic ozone hole because temperatures there are not cold enough to allow the cloud crystals to form.

Ozone is produced in the upper stratosphere, between 20 and 30 miles in altitude, when oxygen molecules absorb solar ultraviolet radiation. When the chemical bond ruptures, it produces ozone, an unstable molecule of three oxygen atoms along with a single oxygen atom. Ozone then decays back to an oxygen molecule and an oxygen atom within the ozone layer. Certain chemicals that are released into the stratosphere directly compete for the free oxygen atoms in the ozone layer and interfere with ozone production. The mechanisms by which ozone is destroyed are based on chemical chain reactions. One pollutant molecule might destroy many thousands of ozone molecules before being transported to the lower atmosphere, where it can no longer do any harm.

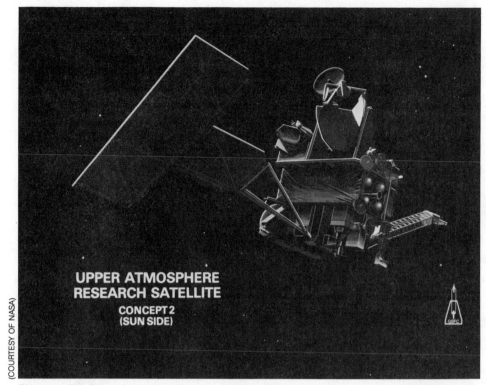

Fig. 6-11. The Upper Atmosphere Research Satellite will measure ozone and other constituents of the upper atmosphere.

The ozone layer is only a trace constituent of the stratosphere, with a maximum concentration of only a few parts per million of the air molecules. If the diffuse ozone layer were concentrated into a thin shell of pure ozone gas surrounding the Earth at atmospheric pressure, it would only measure about an eighth of an inch thick.

The variation of ozone concentrations in the ozone layer is affected by the seasons, latitude, and strong weather systems that penetrate the stratosphere. If ozone finds its way to the lower levels of the atmosphere, it is destroyed before reaching the ground. This is fortunate because ozone also happens to be highly toxic when near the surface. It can irritate the eyes and lungs if it is present in the lower air.

Ozone plays a very important role in shielding the Earth from harmful shortwave, ultraviolet radiation. Without this shield, life could not exist on the Earth's surface. Even a slight increase in ultraviolet rays can cause skin cancer and cataracts, have harmful effects on plants and animals, and exacerbate pollution problems such as smog and acid rain.

Laboratory experiments have shown that certain chemical substances, particularly chlorofluorocarbons (CFCs) and nitrous oxides, destroy ozone. CFCs are used as refrigerants in refrigerators and air conditioners. They escape into the atmosphere when these appliances are manufactured and later discarded. They are used as propellants in spray cans and in the manufacture of foam plastics. CFCs are also used as industrial solvents. Evaporation and spillage sends many of these chemicals into the atmosphere. In addition, CFCs soak up infrared energy and are 10,000 times more effective than carbon dioxide at trapping escaping heat from the Earth. They thereby contribute significantly to greenhouse warming.

Nitrous oxides are produced by the combustion of fossil fuels, especially under high temperatures and pressures like those found in coal-fired plants and internal combustion engines. The tall chimneys of coal-fired plants send huge amounts of nitrous oxides high into the atmosphere where a portion mixes with water and rains out as nitric acid. The remainder finds its way into the upper stratosphere and breaks down ozone. The continued depletion of the ozone layer and the accompanying high ultraviolet exposure could reduce crop productivity and aquatic life. Primary producers, upon which ultimately all life on Earth depends for its survival, will be especially hard hit.

7

Deforestation

T HE world is in danger of losing its forests. Tropical rain forests, especially those in the Amazon Basin of South America are being destroyed at a rate of about 30 million acres every year. Only an area of rain forest roughly the size of the United States remains, and an area the size of Alabama is destroyed annually. If the present global deforestation continues, the world's rain forests and their inhabitants will be nearly gone by the middle of the next century. The forests are disappearing by small-scale slash-and-burn agriculture and large-scale timber harvesting. Some tropical forest fires are so huge they create gigantic smoke plumes that can be viewed from space (FIG. 7-1).

One method of slowing down the carbon dioxide buildup in the atmosphere is to plant more trees. By doubling the volume of forest growth each year, the major fossil fuel consuming nations could delay the onset of greenhouse warming for perhaps a decade or more. The destruction of existing rain forests, however, would have to be halted immediately. It would take a forest covering nearly 3 million square miles, or approximately the size of Australia, to fully restore the Earth's carbon dioxide balance. This is an area equal to all the tropical forests that have been cleared since the advent of agriculture.

WORLD ENVIRONMENTS

The Earth's surface is approximately one-third desert; one-third forests, savannah, and wetlands; and one-fifth glacial ice and tundra (FIG. 7-2). The rest is occupied by humans. Deserts are not only the hottest and driest regions, but they are also among the most barren environments in the world. Only the hardiest plant and animal species can survive in these arid regions. Technically, a desert is defined as a region that receives less than 10 inches of precipitation annually. Much of the world's desert waste-lands receive only a minor amount of rain during certain times of the year. Other areas have gone for years without rain.

Because of natural and human activities, increasing amounts of land are becoming deserts. Some 15,000 square miles of new desert, or an area about the size of California's Mojave desert, are created each year. About 10 percent of the world's deserts are composed of sandy dunes, which are driven across the desert floor by winds (FIG. 7-3).

(COURTESY OF NASA)

Fig. 7-1. The Amazon basin of South America is obscured by smoke from clearing and burning of the tropical rain forest as viewed from the Space Shuttle Discovery *in December 1988.*

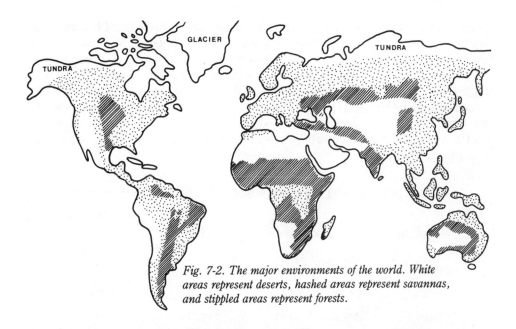

Fig. 7-2. The major environments of the world. White areas represent deserts, hashed areas represent savannas, and stippled areas represent forests.

Fig. 7-3. Large sand dunes in Death Valley, California.

Perhaps the most impoverished desert on the face of the Earth is found in Antarctica. Dry valleys that run between McMurdo Sound and the Transantarctic Mountains receive less than 4 inches of snowfall each year, most of which is blown away by strong winds.

Most of the world's great deserts exist in the subtropics in a broad band that runs roughly between 15 and 40 degrees latitude north and south of the equator (TABLE 7-1). In the Northern Hemisphere, a series of deserts stretches from the west coast of northern Africa through the Arabian peninsula and Iran and into India and China. In the Southern Hemisphere, a band of deserts runs across southern Africa, central Australia, and west-central South America.

TABLE 7-1. Major Deserts of the World

DESERT	LOCATION	TYPE	AREA SQUARE MILES × 1000
Sahara	Northern Africa	Tropical	3500
Australian	Western/interior	Tropical	1300
Arabian	Arabian Peninsula	Tropical	1000
Turkestan	S. Central U.S.S.R.	Continental	750
North America	S.W. U.S./N. Mexico	Continental	500
Patagonian	Argentina	Continental	260
Thar	India/Pakistan	Tropical	230
Kalahari	S.W. Africa	Littoral	220
Gobi	Mongolia/China	Continental	200
Takla Makan	Sinkiang, China	Continental	200
Iranian	Iran/Afganistan	Tropical	150
Atacama	Peru/Chile	Littoral	140

After the rain falls out of the rising tropical air, there is little moisture left for the subtropics. The dry air cools and sinks, which produces zones of semipermanent high pressure. The high pressure produces beautifully clear skies and calm winds most of the time, but also blocks advancing weather systems from entering the region. Tall mountains also block weather systems. Generally, on the leeward side of a mountain there is an area called a rain shadow zone where only a small amount of precipitation falls. This is because the mountains force clouds to rise, which produces precipitation on the windward side and leaves the leeward side dry (FIG. 7-4).

Since deserts are generally light-colored, they have a high albedo and reflect large amounts of solar energy back into space. Desert sands also absorb a great deal of heat during the day. Surface temperatures can exceed 150 degrees Fahrenheit. Because the skies are mostly clear during the night, however, thermal energy that was trapped in the

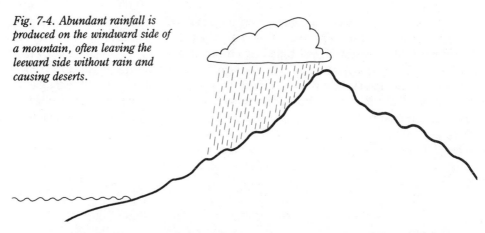

Fig. 7-4. Abundant rainfall is produced on the windward side of a mountain, often leaving the leeward side without rain and causing deserts.

sand during the day quickly escapes. Deserts can be very cold at night, with temperatures sometimes dropping near freezing. This gives desert regions the greatest daily temperature extremes of any place on Earth.

One of the most barren environments is the arctic tundra of North America and Eurasia. Tundra covers about 14 percent of the world's land surface in an irregular band that winds around the top of the world. It exists north of the tree line and south of the permanent ice sheets (FIG. 7-5). Alpine tundra exists in much of the world's mountainous terrain above the tree line and below mountain glaciers. Unlike arctic tundra, which lies at high latitudes and therefore is deprived of sunlight during the long winters, alpine tundra receives sunlight daily. While little snow falls in much of the Arctic, alpine areas receive abundant snowfall because of their high elevation.

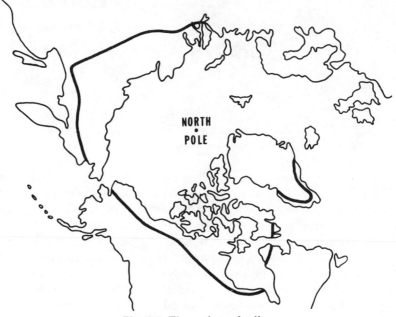

NORTH
•
POLE

Fig. 7-5. The arctic tundra line.

The vegetation in the two regions consists mostly of stunted plants that are often widely separated by bare soil or rock (FIG. 7-6). Most of the ground in the arctic tundra, called permafrost, is frozen year round. Even though the ground is bathed in 24-hour sunlight, the soil temperature seldom rises much above the freezing point. This is because most of the sunlight is used to melt the ice in the soil. As a result, only the top few inches of soil thaws during the short summer season. This often produces patterned ground, which makes the tundra look like a tiled floor (FIG. 7-7). The polygonal patterns are produced by a sequence of freeze-thaw cycles. The soil expands and pushes larger rocks upward and outward. The result is a mound of dirt surrounded by boulders.

(PHOTO BY J.R. WILLIAMS, COURTESY OF USGS)

Fig. 7-6. The arctic tundra in southwestern Copper River Basin, Alaska.

The arctic tundra is one of the most fragile environments in the world. Even small disturbances can cause a great deal of damage. As the polar front swoops across polluted regions of the Northern Hemisphere during the winter, it removes atmospheric pollution and transports it to the arctic regions. There the pollution contaminates the once pristine skies and produces what is called arctic haze. The haze originates mostly in Europe and northwest Asia. At times it is as bad as that of some urban American suburbs.

Wetlands are perhaps the richest of all ecosystems. They produce as much as eight times more plant matter per acre than an average wheat field. A large variety of plants and animals make wetlands their home. Coastal wetlands support valuable fisheries.

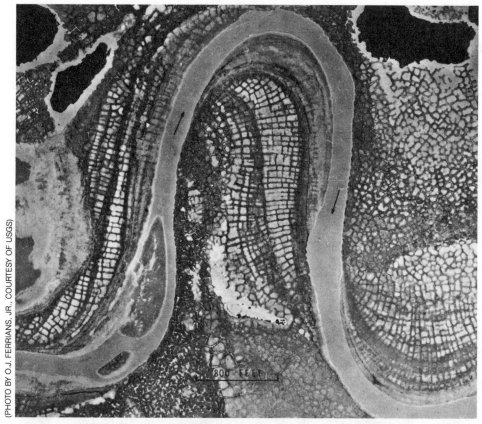

Fig. 7-7. Polygonal markings on the ground surface near Meade River in northern Alaska.

About two-thirds of the shellfish harvested in the U.S. use these areas as spawning and nursery grounds. Wetlands act as a natural filter by removing sediments and water pollution. They also protect coasts from storms and erosion.

Yet, the world's wetlands are disappearing at an alarming rate. As sea levels continue to rise, 80 percent of the U.S. coastal wetlands could be lost by the middle of the next century. Woodland wetlands are disappearing at a rate of about 1,200 acres per day (FIG. 7-8). Almost 90 percent of recent wetland losses in the United States have been for agricultural purposes. Short-term food production is replacing the long-term economic and ecological benefits of the wetlands. The consequences include the loss of local fisheries and breeding grounds for marine species and wildlife. Like rain forests, the destruction of the wetlands in many cases is irreversible.

Tropical rain forests were named because they receive upwards of 200 inches or more of rain each year. They only cover about 6 percent of the land surface, yet contain two-thirds or more of the over 10 million species on Earth. The plants and animals of the rain forests are continuously being crowded out by the encroachment of humans. Their habitat is being destroyed by herbicides, insecticides, and industrial wastes.

Fig. 7-8. The Swan Lake waterfowl refuge in South Dakota dried out in 1974 due to drought.

(PHOTO BY PAT KUCK, COURTESY OF USDA-SOIL CONVERSATION SERVICE)

Some exotic plants that are in danger of becoming extinct have important medicinal value. Over half the pharmaceuticals used today are derived from natural herbs, most of which exist in the rain forests. Sometime during the next century, if the present rate of extinction continues, the number of species lost because of human activities could surpass that of the great die-out at the end of the Cretaceous period when the dinosaurs and three-quarters of all other species disappeared 65 million years ago.

FOREST DESTRUCTION

Half the forests of the world have been cut down to make room for farmland since the dawn of agriculture. Only about 15 percent of the United States' once vast forests remain. Tropical rain forests are being cleared at a rate of about 50,000 square miles every year. Approximately 75 percent of the deforestation is conducted by landless people in a desperate search for food.

Rain forests exist primarily on thin, nutrient-poor soil. Half the rooting depth of a typical Amazon tree can be washed away in its lifetime. When the trees are burned the ashes help fertilize the soil, but after a couple of years of intense agriculture the soil is robbed of its nutrients and farmers are forced to abandon their farms. More trees are then cut down to provide new agricultural land.

The process of deforestation is happening all over the world. It is at its worse, however, in the Amazon jungle of Brazil (FIG. 7-9). Some 20 million acres of forests are destroyed there each year through a process called slash and burn. As much as 20 percent of the Amazon rain forest has been destroyed in this manner. The smoke from burning trees is so thick and far-reaching that it has caused commercial airports in northern Brazil to close because of poor visibility.

Fig. 7-9. The extent of the Amazon rain forest.

After the soil is robbed of its nutrients, the land is abandoned. When the rains come, usually in September, flash floods wash away the denuded soil and expose bedrock. Without the soil, the rain forest has no chance to recover. The forests generate about half the life-supporting precipitation because of the high transpiration rates of the dense vegetation. Precipitation patterns are changing because large parts of the rain forests are being destroyed. The potential outcome is for large areas to turn into man-made deserts.

Even the temperate forests of the higher latitudes are not immune to destruction. Over the past two decades, the growth rate and general health of forests in the northeastern United States, eastern Canada, and many parts of central Europe have been declining. There are a number of factors responsible for this, the worst being acid rain. The acid precipitation runs off into streams and lakes and percolates into the soil. It damages plant roots and leaches out valuable soil nutrients. The direct contact of acids on foliage also destroys trees as well as agricultural crops.

Orbiting satellites have a unique perspective from which to study the forests (FIG. 7-10). Because of the continuous coverage, changes in the forests can be detected. In addition, deforestation rates can be accurately assessed by comparing satellite imagery from different times. Different types of vegetation also have unique special signatures, which can be used for broad-area identification.

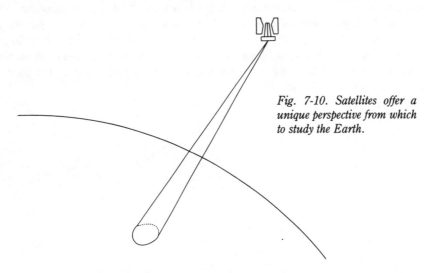

Fig. 7-10. Satellites offer a unique perspective from which to study the Earth.

Vegetation stress generally reveals itself in changes in leaf structure, chlorophyll content, and water content. These changes can be detected with satellite imagery. Healthy vegetation absorbs red light and reflects infrared light, while bare ground reflects red light and absorbs infrared light. When a patch of forest is cut down, the change in satellite imagery is clearly visible. The destroyed area can be measured with considerable accuracy. In addition to mapping and monitoring forest decline and destruction, future satellite systems might provide much needed information about the specific causes of deforestation such as disease, infestation, drought, and human activities.

The biota, which is all living things on the surface of the Earth, and humus, which is dead organic matter in the soil, hold about 40 times more carbon than the entire atmosphere. The harvest of forests, the extension of agriculture, and the destruction of wetlands destroys wildlife habitats and speeds the decay of humus, which sends huge amounts of carbon dioxide into the atmosphere. Agricultural lands, which produce carbon dioxide when cultivated, do not store nearly as much carbon as the forests they replace. The clearing of land for agriculture, especially in the tropics, is the largest source of carbon released into the atmosphere from the biota and soils.

Forests conduct more photosynthesis for a given area than any other form of vegetation. They incorporate from 10 to 20 times more carbon per acre than cropland or pastureland and contain as much carbon as the entire atmosphere. Deforestation presently accounts for about one-fifth the total amount of carbon dioxide and as much as one-half the amount of methane being released into the atmosphere. With the reduction of the forests, stores of carbon in the trees are released into the atmosphere, and the ability of the forests to remove excess atmospheric carbon dioxide is lessened. This can further stimulate greenhouse warming.

MASS EXTINCTIONS

Since the beginning of life, species have come and gone (FIG. 7-11). Species alive today represent less than 1 percent of all the species that have ever lived. It is for this reason that extinctions play a fundamental role in the evolution of life. All extinctions seem to indicate biological systems in extreme stress, which has been brought on by rapid environmental change. When a major extinction takes place, the evolutionary clock is reset and new species develop to take the places that have been vacated.

Fig. 7-11. The rise and fall in the number of species over the past 570 million years. The large dip is in response to the great Permian extinction 240 million years ago when 95 percent of all species went extinct.

The most recent mass extinction occurred at the end of the last ice age about 11,000 years ago. It was a most unusual extinction because it only affected large land mammal species. As the glaciers began to retreat, the global environment adjusted to the changing climate. The result was a shrinking of the forests and an expansion of the grasslands. This disrupted the food chain of the large animals. Deprived of their food resources, they died out. By this time, humans were becoming efficient hunters and may have had a role in the extinction of herds of wildlife.

It appears that nature is constantly experimenting with different forms of life. When one fails, it dies off and is never seen again. Once a species becomes extinct, it is gone forever. The odds of that special gene combination reappearing are astronomical. This seems to put evolution on a one-way track. Although it perfects species to live at their optimum in their respective environments, it can never go back to the past. From the very beginning, life has advanced from the simple to the complex. It has not been convincingly demonstrated that a species has ever degenerated to a lower form.

The extinctions of the past were caused by natural phenomena. Present-day extinctions, which amount to about a hundred species per day, are caused by human activities. If the present spiral of human and environmental destruction continues out of control, half or more of all species living on Earth today will be gone, possibly as early as the middle of the next century.

This places man in the unique position of being the only animal to have the ability to affect the extinctions of large numbers of other species. The complex interrelationships among species and between them and their environments is still not fully understood. It is becoming more apparent, however, that the destruction of large numbers of species will leave this an entirely different biological world.

CLIMATIC EFFECTS

Forests exert a strong control on the climate. The loss of forests increases surface albedo, so more sunlight is reflected into space. This loss of solar energy could change precipitation patterns with a consequential decrease in rainfall, particularly in the rain forests. This change could cause stress and make the trees more susceptible to disease. Soot from forest fires absorbs sunlight, which heats the atmosphere. This produces a temperature imbalance and makes the temperature increase with altitude, which is just the reverse of what it should do. Large quantities of soot in the atmosphere generated by massive forest fires could therefore send abnormal weather patterns around the world.

Forests also store a great deal of carbon. The clearing of forested land for agriculture, especially in the tropics, is one of the largest sources of released carbon. Although there is a net accumulation of carbon in the forests of North America and Europe, it is insignificant compared to the losses in the tropical regions. The increased carbon dioxide content in the atmosphere heightens the greenhouse effect, which can substantially change global weather patterns.

Deforestation might also be a threat to the ozone layer. Clear-cutting of timber encourages soil bacteria to produce nitrous oxide, which is released into the air. The tremendous heat produced by the burning trees combines nitrogen with oxygen to make nitrous oxide. If significant amounts of this gas enter the upper atmosphere, it will destroy the ozone layer.

Methane is the second most important greenhouse gas. The atmosphere presently contains about 1 molecule of methane for every 200 molecules of carbon dioxide. Methane is transparent to certain wavelengths of light that carbon dioxide is not. It is also 20 to 30 times more effective per molecule at absorbing infrared radiation than carbon dioxide. This means that even small amounts of methane can have a large effect. Methane production is outstripping that of carbon dioxide. It is increasing at a rate of about 1 percent per year compared to a rate of about one-half of 1 percent for carbon dioxide. In the ensuing years, methane and other trace greenhouse gasses might together contribute more to greenhouse warming than carbon dioxide.

Deforestation is increasing rapidly the number of termites. Presently, there are about 1,500 pounds of termites for every person on Earth. As deforestation escalates, that number could increase by a factor of 10. Termites ingest as much as two-thirds of all the carbon on land. About 1 percent of this carbon is converted into methane. In addition, large numbers of cattle that are raised on cleared land contribute substantial amounts of atmospheric methane from their digestive process. With one bovine for every four people on Earth, these animals might play a substantial role in changing the climate of the planet.

REFORESTATION

One of the ways to combat the greenhouse effect is to plant more trees. Additional trees would absorb the excess carbon dioxide in the atmosphere. The replanting of 100 million trees would remove about 18 million tons of carbon dioxide from the atmosphere each year. In the United States, lumber companies reseed the forests they cut down to preserve a future supply of forest products. Young trees, however, do not absorb as much carbon dioxide as the older trees they replaced. Even the fastest growing tree, dubbed the super pine, still takes 20 to 30 years to mature.

(COURTESY OF NATIONAL PARK SERVICE)

Fig. 7-12. Forest fires in Yellowstone National Park destroyed nearly half the forested land in the summer and fall of 1988.

For much of the world, deforestation has destroyed the topsoil to such an extent that it is no longer possible to replant trees. There are plenty of degraded lands around the world, however, that could be replanted in forests without conflicting with agriculture.

Nations are beginning to set aside forested areas in an attempt to halt the tide of deforestation and to preserve animals like the elephant and black rhinoceros. The amount of forested land in the United States and a few other countries has actually increased slightly in recent years. The United States Forest Service has taken millions of acres of forested land out of multiple use and set up wilderness areas. Forests surrounding those wilderness areas, however, are still being exploited to their fullest.

As the Earth heats up and the forests dry out, extensive forest fires could lay waste to huge tracks of valuable forested land as they did at Yellowstone National Park during the summer and fall of 1988 (FIG. 7-12). If global temperatures rise too rapidly, the forests will not be able to keep up with the movement of climate zones into higher latitudes, which could cause a further decline in the world's forests.

8

Agriculture

IT is becoming apparent that our planet is constantly changing. Continents move about, seas fill and dry out, mountains rise and are eroded down to plains, glaciers expand and retreat, and species come and go. Some 450 million years ago, before plants began to seed the landscape, the Earth was a desert planet. At no other time since then has the Earth been covered mostly by deserts.

THE FOOD REVOLUTION

When the last ice age ended and the glaciers retreated from the northern lands, the most dramatic climate changes in the history of the planet occurred. Beginning about 9,000 years ago, there was a long wet spell that was caused by the warming of the interior of the continents. This strengthened the monsoon winds, which brought moisture-laden sea breezes inland over Africa, India, and southeast Asia. A period known as the Climatic Optimum, which began about 6,000 years ago, saw unusually warm, wet conditions that lasted for 2 thousand years.

These optimum climatic conditions took place during a period known as the Neolithic, or new stone, age. This was also the beginning of a food-producing revolution in the Old World. This same agricultural expansion took place a few thousand years later in the New World, where records show maize was cultivated in the Amazon region of South American around 3,000 years ago.

Even in its earliest stages, agriculture was so productive it could support several times more people in a given area than hunting and gathering. Agriculture both supported and encouraged population growth because it was labor intensive. Large families were required to till and harvest. Increasing populations, however, required food production to intensify, often with disastrous consequences to the land. When resources were exhausted, entire communities collapsed and people were forced to move.

Farmers roamed across Europe and followed the tracks of the hunter-gatherers, who had followed the retreating glaciers northward (FIG. 8-1). About 6,000 B.C., agriculture began to spread into Asia and Europe. It reached northern Europe and Great Britain about 3,000 years later.

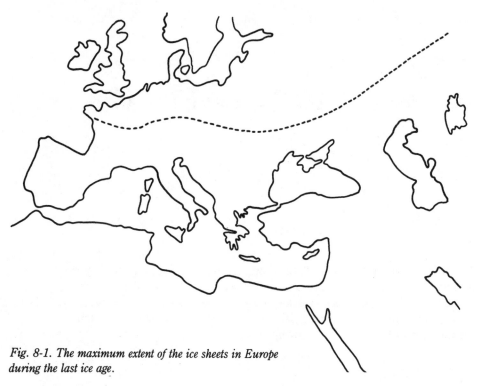

Fig. 8-1. *The maximum extent of the ice sheets in Europe during the last ice age.*

SLASH AND BURN

The once lush Fertile Crescent, which fed as many as 25 million people remains infertile because of over-irrigation and salt accumulation in the soil from Sumerian farmers 6,000 years ago. Over 1,000 years ago, nomads of the Sahel region of central Africa lived by hunting and herding. The Sahel, which is a 250-mile strip of land south of the Sahara desert that extends from the west coast of the continent to Chad (FIG. 8-2), was mostly tropical forest at that time. The nomads cut and burned the trees to improve grazing in the area. This turned a natural forested habitat into a grassland. Overgrazing weakened the soil, and an extensive man-made desert was created.

Today, the Sahel is being overrun by the sands of the Sahara desert, which are engulfing everything in their path. Because there is no vegetation, the land is subjected to flash floods, higher evaporation rates, and tremendous dust storms. The denuded land also has a higher albedo, which contributes to lower rainfall levels and denudes more

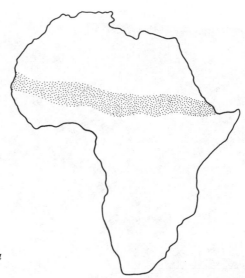

Fig. 8-2. The Sahel region of Africa, which borders the Sahara Desert to the north.

land. Thus, man-made deserts march across once fertile lands. Throughout the world, perhaps as much as one-third to one-half of what was once arable land is now rendered useless by erosion. In addition, one-half to three-quarters of all irrigated land will be destroyed by the accumulation of salt in the soil by the end of this century.

Global rain forests covered an area equal to about twice the size of Europe 100 years ago. That amount of forested land has been cut in half to make room for new agricultural fields. Much of the land has been cleared by slash and burn methods. Trees are set ablaze and their ashes are used to fertilize the thin, nutrient-poor soil. After a year or two of improper farming or grazing, the soil is worn out and farmers are forced to move. The abandoned farms are subjected to severe soil erosion because there is no vegetative cover to protect against the effects of wind and rain. The severely eroded soil can no longer support life.

SOIL EROSION

Perhaps the greatest limiting factor to further human population growth is soil erosion (FIG. 8-3). Before the advent of agriculture, natural soil erosion rates were probably no more than 10 billion tons per year. This level was slow enough that it could be replaced by the generation of new soil. Estimates of present soil erosion rates are about 25 billion tons per year. In other words, we are losing soil more than twice as fast as nature is putting it back. As much as one-third of the world's cropland is losing soil at a rate that is undermining any long-term agricultural productivity. World food production per capita will eventually begin to fall off if the loss of topsoil continues at the present rate.

Erosion rates vary depending on the amount of precipitation, the topography, the type of rock and soil, and the amount of vegetative cover. Efforts to increase worldwide crop production through deforestation, irrigation, use of artificial fertilizers, genetic engineering, and improved farming techniques will ultimately be defeated if the top soil is

Fig. 8-3. More than 6 billion tons of soil erode from the nation's farmlands and other lands each year.

lost. In the United States, eroding cropland pollutes rivers and lakes. The added sediments in these water sources severely limit the life expectancy of dams erected for water projects at a cost of nearly a billion dollars annually. The best way to control silt buildup is to adopt effective soil-conservation measures so that less topsoil is lost to erosion.

The soil profile (FIG. 8-4) is divided into the A and B horizons. The A horizon is a thin zone of soil from a few inches to a few feet thick, with an average thickness of seven inches worldwide. Most soil nutrients are within this zone. The poor soil of the B horizon lies below the A horizon. As the top zone thins out and erosion brings the B horizon to the surface, the potential for runoff and erosion is increased. The B horizon is generally unable to sustain vegetation (TABLE 8-1).

Most of the world's arable land is already under cultivation. Efforts to cultivate substandard soils are leading to poor productivity and ultimately abandonment, which in turn leads to severe soil erosion. Marginal lands, which are often hilly, dry, or contain only thin, fragile topsoils are also being forced into production. As world populations continue to grow and require more food from dwindling resources, man-made deserts will continue to encroach on large parts of the land.

IRRIGATION

Over 10 percent of the world's cultivated land must be irrigated (FIG. 8-5). Some 600 cubic miles of water is used annually. Irrigation has many advantages. Crops need not rely on undependable rainfall, more land can be brought under cultivation, and two or

Fig. 8-4. The soil profile showing the dark, organic-rich topsoil and the sandy, infertile subsoil below. Measurements are marked off in feet.

more crops can be grown in a single year. Irrigation also has its drawbacks. Most river water that is used for irrigation has a high salt content. If fields are not drained properly, the salt buildup in the soil can ruin the land. Tens of thousands of acres of once fertile land are destroyed by this process annually.

California produces about 10 percent of the nation's farm income. By the turn of the century, at least one-third of California's farmland, about 1.5 million acres, could be destroyed by salt. In fact, estimates indicate that by the end of the century, over half of all irrigated land will be rendered useless due to salt accumulation (FIG. 8-6).

Lower water availability in the spring and summer could also dramatically reduce crops. A reduction in the supply of water slows the drainage of agricultural chemicals as well as naturally occurring selenium, arsenic, boron, and other poisons. Selenium leaching from soil on irrigated farms in the San Joaquin Valley of California (FIG. 8-7), has caused deformities in nearby waterfowl.

China has the most irrigated land of any country in the world. Water is provided by some 100,000 dams and reservoirs, which have a total storage capacity of about 100 cubic miles of water. The Nile River serves one of the largest irrigated areas in the world (FIG. 8-8): some 20,000 square miles, or an area about the size of West Virginia. On the border between Zambia and Zimbabwe, the Zambezi River has been dammed to create

TABLE 8-1. Summary of Soil Types

CLIMATE	TEMPERATE (HUMID) > 160 IN. RAINFALL	TEMPERATE (DRY) < 160 IN. RAINFALL	TROPICAL (HEAVY RAINFALL)	ARCTIC OR DESERT
Vegetation	Forest	Grass and brush	Grass and trees	Almost none, no humus development
Typical area	Eastern U.S.	Western U.S.		
Soil type Topsoil	Pedalfer Sandy, light colored; acid	Pedocal Enriched in calcite; white color	Laterite Enriched in iron and aluminum, brick red color	No real soil forms because no organic material. Chemical weathering very low
Subsoil	Enriched in aluminum, iron, and clay; brown color	Enriched in calcite; white color	All other elements removed by leaching	
Remarks	Extreme development in conifer forest abundant humus makes groundwater acid. Soil light gray due to lack of iron	Caliche - name applied to accumulation of calcite	Apparently bacteria destroy humus, no acid available to remove iron	

Kariba Lake, the largest artificial reservoir in the world. Africa, which is prone to drought, depends heavily on its water projects for irrigation. All the large rivers in southern and southeastern Asia have also been developed extensively. This gives Asia the largest volume of impounded water, which is used mostly for irrigation.

The use of groundwater for irrigation is expensive. Generally, only affluent nations can afford it on a large scale. The water is pumped from wells, which in some cases are thousands of feet deep. The overuse of ground water can lower the water table or totally deplete the aquifer. When an aquifer becomes depleted, subsidence causes compaction, which decreases the pore spaces between sediment grains. When this happens, the groundwater system is no longer able to carry its original capacity and wells go dry.

Fig. 8-5a. Irrigated cotton in Imperial County, California.

Fig. 8-5b. Heavily irrigated areas in the U.S.

Fig. 8-6. Areas affected by salt buildup in the soils of the U.S.

DROUGHT

Droughts occur when precipitation patterns shift. Since the total heat budget of the Earth does not change significantly from one year to the next, areas that become unusually dry are matched to some extent by areas that become exceptionally wet. For example, in the 1980s the United States has had a series of bad droughts similar to those of

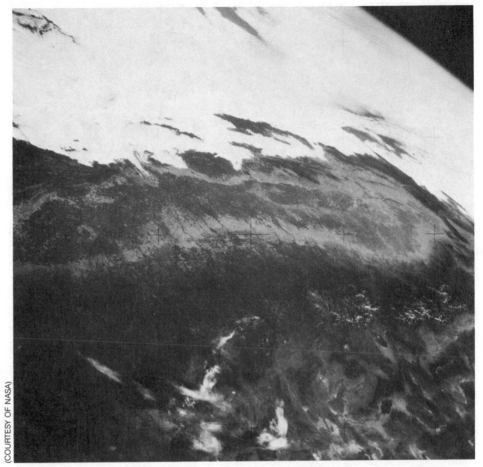

Fig. 8-7. The Sacramento and San Joaquin Valleys of California as viewed from Skylab.

the Dust Bowl years of the 1930s (FIG. 8-9). The U.S. has also experienced abnormally high temperatures over the same period. In 1983, Australia had its most severe drought in over 100 years. An equally intense drought caused food shortages in southern Africa and affected western Africa and the Sahel region. Meanwhile, the worst flooding of the century struck Ecuador, northern Peru, and large areas of Brazil, Paraguay, and Argentina.

The sub-saharan drought of the last quarter century has been the worst in 150 years. The 1983 and 1984 droughts, which left upwards of a million people dead or dying from starvation, were the worst in this century. One contributing factor to such a long-lasting drought could have been the denuding of the land. Stripping away vegetation alters the reflective properties of the soil. Another influence might have been the increase in atmospheric carbon dioxide from the combustion of fossil fuels, the destruction of forests and wetlands, and the extension of agriculture, all of which release large quantities of stored carbon into the atmosphere. The loss of vegetation also reduces the amount of carbon dioxide that is taken out of the atmosphere by photosynthesis.

Fig. 8-8. The Nile River Delta, viewed from the space shuttle, serves some 45 million people in a 7,500-square-mile area.

Fig. 8-9. A dust storm on a farmstead near Elkhart, Kansas, during the 1930s dustbowl.

The accumulation of carbon dioxide in the atmosphere over this century is expected to cause a general warming of the Earth. The increase in global temperatures could dramatically affect the climate and shift precipitation patterns around the world. Wet conditions could come to some areas and droughts to others. Greenhouse warming is also likely to increase both the frequency and severity of droughts.

The central portions of the continents that normally experience occasional drought (FIG. 8-10) could become permanent dry wastelands. Almost all the soil in Europe, Asia, and North America will become drier, and require upwards of 50 percent more irrigation. Despite expected rises in temperatures, increased evaporation, and changes in rainfall patterns, the United States should still be able to produce enough food to feed itself. The country's ability to produce excess food for export, however, could be adversely affected.

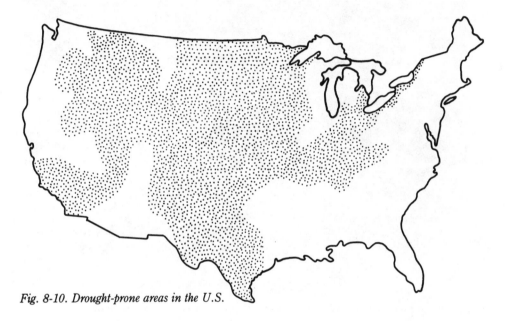

Fig. 8-10. Drought-prone areas in the U.S.

Regions that sit 30 degrees on either side of the equator can expect large changes in precipitation patterns as the world continues to warm. There are also historical variations in annual precipitation, which are normally expected about 30 times each century.

DESERTIFICATION

Desertification is the process of degrading the environment. It is a product of climate and human activity and results from the loss of topsoil. After the land is denuded of the clays and fine silts that constitute the topsoil, only the coarse sands are left behind. If this involves a sizable area, a man-made desert is created. The problem is made worse when the land is subjected to flash floods, higher erosion rates, and dust storms, all of which sweep the sands from one place to another.

Desertification is happening all over the world, but it is most prevalent in central Africa. There the burning sands of the Sahara desert (FIG. 8-11) are marching steadily

Fig. 8-11. View of the Sahara Desert of Central Algeria from the space shuttle.

across what was once fertile farmland. The process of desertification is also self perpetuating. Bright sand reflects sunlight, which produces high-pressure regions that block out weather systems and contribute to lower levels of rainfall. Denuded land is also subjected to flash floods and dust storms, which transport the sediments out of the region. As much as one-third to one-half of the world's once fertile lands have been rendered useless by erosion and desertification.

The denuding of the world's forests increases the albedo with a consequential loss of precipitation. This puts an additional strain on the forests and subjects them to infestation and disease, which causes additional trees to die. In areas where the forests are being destroyed and the ground denuded of vegetation, the process of erosion worsens the situation by exposing infertile soil and bedrock that do not support plant growth.

Desertification cannot fail to have a detrimental effect on the Earth's heat budget, which stabilizes the overall climatic temperature. An unstable atmosphere will produce more turbulence and vicious storms will march across the globe. Attempts to produce additional rainfall or to control nature's rages using weather modification schemes (FIG. 8-12) might in the long run prove insufficient.

Fig. 8-12. Cloud seeding using streams of silver iodide trailing from beneath an aircraft.

OVERPOPULATION

Two million years ago, our ancient ancestors numbered about 100,000 people. When agriculture was invented, roughly 10,000 years ago, there were upwards of 10 million people on the planet. By the time of the first dynasty in Egypt, around 5,000 years ago, the population had increased to 100 million. At the height of the Roman Empire during the first century A.D., the world's population more than doubled. It doubled again at the beginning of the Industrial Revolution and reached its first billion mark around 1800. Now, less than two centuries later, the number of people in the world has increased five fold and is expected to double that amount by the middle of the next century.

Many demographers believe that the world has already reached its carrying capacity, which is the ability of the land to support people. Rapid population growth has stretched the resources of the world. The prospect of future population increases raises serious doubts about whether the planet will be able to continue to support people's growing needs.

When agriculture can no longer supply the necessary food, people will be in grave danger. Mass starvation could occur whenever drought, infestation, or disease results in greatly reduced crop yields (FIG. 8-13).

The leading food exporters already have most of their arable lands in production. During the 1970s, American farmers placed an additional 60 million acres, an area larger than the state of Kansas, under cultivation in order to help feed a hungry world. Much of this increased land was substandard and included sloping, marginal, and fragile soils that erode easily. Because of increasing pressure for more food production, normally fallow fields are being cultivated, which quickly wears out the soil. Attempts to farm the weak soils of the rain forests have proved inadequate. Over-irrigation is also destroying large quantities of acreage due to salt buildup in the soil.

Fig. 8-13. An ear of corn that never reached maturity due to severe drought in the South during the summer of 1986.

(PHOTO BY JUNE DAVIDEK, COURTESY OF USDA)

The total quantity of food directly and indirectly consumed by the human population is staggering. It amounts to roughly a ton per person per year, or about 5 billion tons annually. Nearly half of the total tonnage of crops and three-quarters of the energy and protein content is supplied by cereal grains. Ninety percent of all the world's food comes from only about a dozen crops, most strains of which are genetically undiversified. This means that disease and infestation targeted at those specific strains could wipe out a nation's entire harvest. Also, a large fraction of these grains is eaten by domestic animals, which consume 4 to 7 calories of grain for every calorie of meat they produce.

The average individual food intake is 1,800 calories per day for developing countries and 2,700 calories per day for developed nations. The diets of the poorer countries, however, are not nearly as nutritious as those of the richer nations. The diet of the poorest 20 percent of the world's population falls below the body's requirement for a normally active, healthy life.

The invention of agriculture was perhaps the greatest achievement and possibly the worst mistake in human history. Although it freed people from the constant search for food, it also encouraged them to settle permanently in one area. As a result, man is extremely vulnerable to fluctuations in soil conditions and the climate.

9

The Melting
Ice Caps

ICE covers approximately 10 percent of the planet's surface. About three-quarters of the world's fresh water is locked up in glacial ice. Alpine glaciers, which are found on every continent, hold as much fresh water as all the world's rivers and lakes. During the height of the last ice age, about 18,000 years ago, so much of the Earth's water was locked up in continental ice sheets that sea levels were about 350 feet lower than they are today. Because of the lower seas, shorelines extended outward for several tens of miles. Patches of dry land appeared above the sea, which aided the migration of species from one continent to another.

If the ice caps were to melt entirely, the ocean would rise some 300 feet above its present level. The additional seawater would move shorelines inland 70 miles or more in most places, which would radically change the shapes of the continents. Even the small amount of melting that is presently taking place could have lasting repercussions. We are getting a preview of things to come as beaches and barrier islands slowly disappear and the rising sea begins to drown coastal areas. Without their protective barriers, seashores will be battered continually by raging storms. The rising seas could cause prime agricultural land and valuable wetlands to decrease during a time when they are needed most.

THE ICE AGES

The level of atmospheric carbon dioxide very likely played a major role in the appearance and disappearance of all the ice ages. The Soviet Vostok ice core taken from 7,000 feet beneath eastern Antarctica contains a continuous record of the temperature and composition of the atmosphere for the past 160,000 years. This covers a period that includes the present interglacial period, the last ice age, and the warm interglacial prior to it. Air bubbles trapped in the ice core provide information on the atmospheric carbon dioxide content at the time the ice was laid down. The presence of deuterium, an isotope of hydrogen, provides data on the temperature. The data indicate that the level of carbon dioxide and the temperature kept pace throughout the entire period (FIG. 9-1). The level of atmospheric carbon dioxide during the last ice age was about 0.02 percent, which is about half its present value.

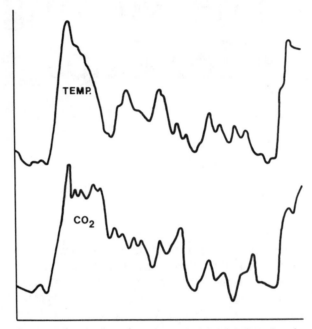

Fig. 9-1. *Global temperatures and atmospheric carbon dioxide levels have kept in step for the past 160,000 years.*

Atmospheric methane, which is the second most important greenhouse gas also fluctuated during the last ice age. About 22,000 years ago, at the height of the glaciation, the level of methane was roughly half of its preindustrial level. The reduction of methane was probably due to the diminished biological activity of the wetlands and other habitats because of the colder climate. During the present interglacial, methane has more than doubled, mainly due to man's activities.

The loss of greenhouse gasses, principally carbon dioxide, might have cooled the climate sufficiently to produce the first known ice age in geological history, about 2 billion years ago. Another glacial episode that occurred some 260 million years ago might have been triggered by the spread of forests on the land as plants adapted to living and reproducing out of the sea. The Earth cooled when the forests removed atmospheric carbon dioxide and converted it into organic matter, which was made into coal and buried in the Earth's crust.

The burial of large amounts of carbon might have been the key to the onset of perhaps the greatest of all ice ages during the Late Precambrian era about 700 million years ago. The glaciations of the Late Ordovician period around 440 million years ago, the Middle Carboniferous period around 330 million years ago, the Permo-Carboniferous period around 290 million years ago, as well as the most recent glacial epoch that began about 2 million years ago, also might have been triggered when atmospheric carbon dioxide was reduced to a level of about one-quarter of its present value.

Only recently have atmospheric scientists acquired enough information on global geochemical cycles to understand what might have caused such a change in the concentration of carbon dioxide. New data taken from deep-sea cores indicate that carbon dioxide variations preceded changes in the extent of the more recent ice sheets. It is expected that the earlier glacial epochs were similarly affected. The variations of carbon dioxide levels might not be the sole cause of glaciation. When combined with other processes, however, they could have been a strong influence.

The positions of the continents also might have substantially influenced the initiation of the ice ages. In the Earth's early history continents were confined to a region around the equator (FIG. 9-2), and the polar regions were vast open bodies of water. Then the continents broke apart and began to move to higher latitudes. Continental movements are thought to be responsible mainly for the Late Ordovician ice age around 440 million years ago. The study of magnetic orientations in rocks from many parts of the world indicates that northern Africa was directly over the South Pole during the Ordovician period, which would have allowed massive ice sheets to grow.

Global tectonics might also have triggered the ice ages because of volcanic activity and seafloor spreading, which drew oxygen out of the ocean and atmosphere so that

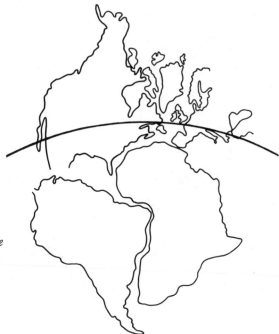

Fig. 9-2. The approximate positions of the continents relative to the equator during the Carboniferous period.

more organic carbon was preserved in the Earth's crust and not returned to the atmosphere.

The last glacial period took place during the Pleistocene epoch, which began about 2.4 million years ago. This was a time when a progression of ice ages came and went, with each being followed by a short, warm interglacial period (FIG. 9-3) similar to the one we are experiencing now. The last ice age began about 100,000 years ago, intensified 75,000 years ago, peaked about 18,000 years ago, and retreated about 10,000 years ago.

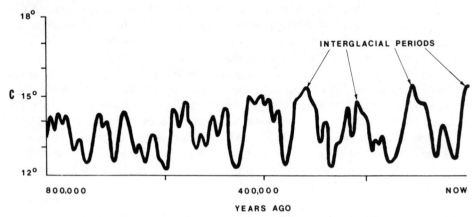

Fig. 9-3. The ice ages were followed by short interglacial periods.

Fig. 9-4. The extended shoreline during the height of the last ice age.

At the height of the ice age, about 5 percent of the planet's water was locked up in glacial ice, which resulted in an appreciable lowering of the sea level and the expansion of the land area by 8 percent (FIG. 9-4).

THE SHRINKING ICE

After some 90,000 years of gradual accumulation, snow and ice was up to 2 miles thick in parts of North America and Eurasia during the last ice age (FIG. 9-5). Yet, the glaciers melted away in only a matter of a few thousand years, retreating upwards of 2,000 feet a year. At least one-third of the ice melted between 16,000 and 13,000 years ago, when average global temperatures increased about 10 degrees Fahrenheit to nearly present-day values. As the North American ice sheet began to retreat, its meltwater flowed down the Mississippi River and into the Gulf of Mexico. After the ice sheet retreated beyond the Great lakes, however, the meltwater ran down the St. Lawrence River (FIG. 9-6), and into the North Atlantic Ocean.

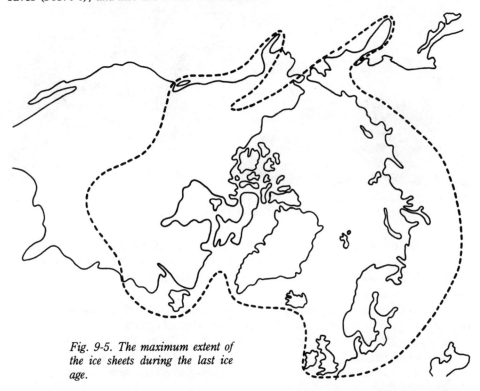

Fig. 9-5. The maximum extent of the ice sheets during the last ice age.

The rapid melting of the glaciers sent a torrent of meltwater and icebergs into the North Atlantic. The cold water formed a freshwater lid on top of the ocean and significantly changed the salinity of the seawater. The cold waters also blocked poleward flowing warm currents from the tropics and caused temperatures on the land to return to near ice age levels. As a result, the ice sheets appeared to pause in mid stride between 11,500 and 10,700 years ago, during a period known as the Younger Dryas.

Afterwards, the warm currents were reestablished and the warming trend re-

Fig. 9-6. The ice-jammed St. Lawrence Seaway in Quebec Province, Canada, just as it might have been when the great ice sheets melted at the end of the last ice age.

mained. This climate change took less than two decades and prompted a second episode of melting. The ice retreated quickly to the north until the present volume of ice was reached about 6,000 years ago. Then began the Climatic Optimum, a period of unusually warm and wet conditions that lasted for 2,000 years.

In North America, the glaciers radically changed atmospheric circulation, which in turn affected storm patterns over the continent. High-pressure centers over the ice sheet brought strong easterly winds along the southern flank of the glacier and strengthened the jet stream aloft. This made it possible for the Mojave and nearby deserts of the southwestern United States to receive enough rainfall to sustain woodlands after the ice sheets began their retreat.

From about 9,000 to 6,000 years ago, when the glaciers over North America shrank to nearly their present position in the Arctic, precipitation over much of the Midwest dropped by as much as 25 percent. At the same time, mean July temperatures rose by as much as 4 degrees Fahrenheit. It also appears that the postglacial eastern and southeastern United States were not significantly warmer 6,000 years ago than they are today. This is probably because the Atlantic Gulf Stream current was reestablished at that time, which moderated the climate.

Between 6,000 and 4,000 years ago, during the Climatic Optimum, many regions of the world warmed by an average of 5 degrees Fahrenheit. The melting ice caps released a torrent of floodwater into the ocean and raised sea levels to nearly their present value, which is 300 feet above where they were when the ice sheets began to melt.

THE POLAR ICE CAPS

About 60 million years ago, the Atlantic Ocean widened until it separated the continents and closed off the Arctic basin from warm, tropical currents. As a result, pack ice accumulated in the Arctic waters. Until then, the Earth seldom had permanent ice caps.

Antarctica became a continent of ice about 40 million years ago, when it detached from Australia and South America and moved into the South Polar region. With the establishment of a circum-Antarctic ocean current, Antarctica became isolated from warm-water currents that originated in the tropics. Ice sheets then started to spread over the eastern end of the continent. About 13 million years ago, the present ice sheet formed as the climate grew colder. During its stay over the South Pole, Antarctica has frozen and thawed several times as evidenced by rising and falling sea levels and the fossilized plants that have been discovered in the interior of the continent.

Three percent of the Earth's water is locked up in the polar ice caps. The ice caps cover, on average, 7 percent of the Earth's surface area. The Arctic is a sea of pack ice (FIG. 9-7), which covers an area of about 4 million square miles and has an average thickness of several tens of feet. If the entire Arctic pack ice were to melt, it would raise global sea levels perhaps only a few feet. If all the ice at the opposite pole in Antarctica (FIG. 9-8) were to melt, however, sea levels would rise by as much as 300 feet.

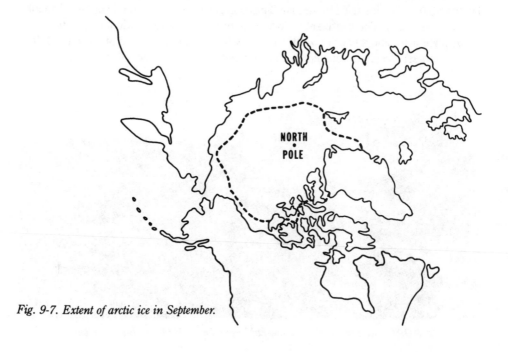

Fig. 9-7. Extent of arctic ice in September.

Fig. 9-8. The Antarctic Peninsula ice plateau showing mountains literally buried in ice.

Nearly 90 percent of all the ice in the world lies on top of Antarctica. With an area of 5.5 million square miles, this desolate land of ice is one and a half times larger than the United States. The ice rises more than 2 miles in places and has a total volume of about 7 million cubic miles. The continent is divided by the Transantarctic Mountains into a large eastern ice mass and a smaller western lobe, which is about the size of Greenland. The world's largest island, Greenland, is also covered by a thick sheet of ice (FIG. 9-9).

Fig. 9-9. The ice-covered Rasmussen Lake near Thule Air Force Base, Greenland.

During the winter months from June to September, nearly 8 million square miles of the ocean that surrounds Antarctica is covered by sea ice with an average thickness of less than 3 feet (FIG. 9-10). Because of this immense expanse of ice, Antarctica plays a greater role in atmospheric and oceanic circulation (FIG. 9-11) than the Arctic. The sea ice is punctured in various spots by coastal and open-ocean polynyas, which are large open-water areas kept from freezing by the upwelling of warmer water from below. These open bodies of water release a tremendous amount of heat, which can rise high into the stratosphere. The upwelling water also has a high carbon dioxide content and could release large enough quantities of this gas into the atmosphere to add significantly to global greenhouse warming.

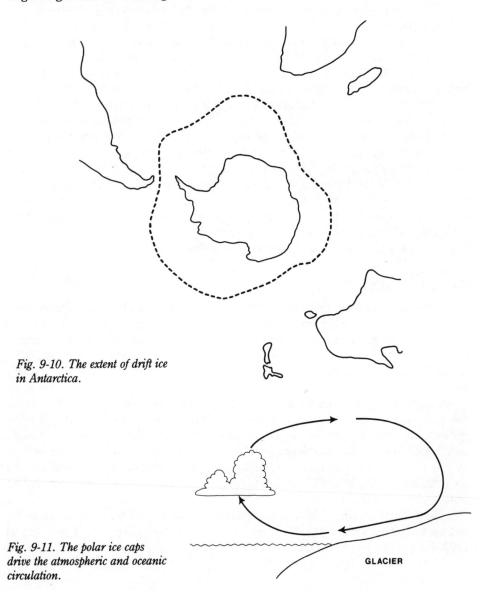

Fig. 9-10. The extent of drift ice in Antarctica.

Fig. 9-11. The polar ice caps drive the atmospheric and oceanic circulation.

GLACIER

Antarctica dumps over a trillion tons of ice into the surrounding seas each year. The ice flowing into the ocean breaks off into icebergs, which appear to be getting larger. The number of extremely large icebergs has also increased dramatically. The largest one found so far measured 100 miles long, 25 miles wide, and 750 feet thick. In August 1989, this iceberg collided with Antarctica and broke in two.

The ice in eastern Antarctica is firmly anchored on land. The ice in the western lobe, however, rests below the sea on shelves of bedrock. It is then surrounded by floating ice that is pinned in by small islands that are buried under the ice. Western Antarctica is traversed by rivers of solid ice that are several miles broad. These ice streams flow down the mountain valleys to the sea and on to the great ice shelves of the Ross and Weddell seas. The banks and midsections of the ice streams are marked by deep crevasses. The bottom of the ice streams might contain muddy pools of melted water, which help lubricate the glaciers and allow them to glide along on the valley floor.

A warmer climate, generated by an increase in atmospheric carbon dioxide, could cause the unstable western Antarctic ice sheet to break loose and crash into the sea. This would raise global sea levels about 20 feet and inundate coastal areas. Even a slow melting of both polar ice caps could raise the level of the oceans upwards of 12 feet by the end of the next century. As a result, much of the world's coastal plains and coastal cities would drown. A rise in the level of the sea could also lift the western Antarctic ice shelves off the seafloor and set them adrift into warm equatorial waters. There they would rapidly melt and raise the sea still higher.

A flood of ice could then surge into the southern ocean, raise sea levels still higher, and set more ice free. The increased area of ice could form a gigantic ice shelf, which could cover as much as 10 million square miles. The increased area of ice would increase the Earth's albedo, which in turn could substantially cool the climate and cause severe instabilities in atmospheric and oceanic circulation systems.

GLACIAL SURGE

Today there are about 200 surge glaciers in North America (FIG. 9-12). During most of its life, a surge glacier behaves normally. It moves along at a snail's pace of perhaps a couple inches a day. At regular intervals of 10 to 100 years, however, the glacier gallops forward up to 100 times faster than normal. A dramatic example is the Bruarjokull Glacier in Iceland. In a single year it advanced 5 miles at an astonishing speed of 16 feet per hour.

A surge often progresses along a glacier like a great wave, proceeding from one section to the next. Subglacial streams of meltwater might act as a lubricant and allow the glacier to flow rapidly. There is no good explanation as to why glaciers surge. They might be influenced by the climate, volcanic heat, and/or earthquakes. However, surge glaciers exist in the same areas as normal glaciers, often almost side by side. In addition, the great Alaskan earthquake of 1964 failed to produce more surges than there were before (FIG. 9-13).

About 800 years ago, Hubbard Glacier in Alaska charged toward the sea, retreated, and then advanced again 500 years later. The last time the glacier surged was around 1900. For the past 85 years, Hubbard Glacier has been flowing toward the Gulf of Alaska

Fig. 9-12. Tikke Glacier is one of nearly 200 surge glaciers in Alaska and adjacent Canada.

at a steady rate of about 200 feet per year. In June 1986, however, the 80-mile-long river of ice surged ahead as much as 46 feet in a single day. There are 20 such glaciers around the Gulf of Alaska that are headed toward the sea. If they should reach the ocean and raise the sea level, the western Antarctic ice shelves could rise off the seafloor and be set adrift. This would cause the level of the sea to rise even higher, which in turn would release more ice, and set in motion a vicious circle.

RISING SEA LEVELS

The level of the sea is definitely rising. The rate seems to be increasing as much as 10 times faster than it was 40 years ago. In most temperate and tropical regions, the sea level is now rising about a quarter of an inch a year. Most of the increase appears to be due to the melting of the western Antarctic and Greenland ice sheets, which is the result of over a decade of global warming. The increase in temperature also causes a thermal expansion of the ocean, which increases its overall volume. Over the last century, thermal expansion has raised the level of the sea by about 2 inches. Alpine glaciers, which contain substantial amounts of the world's ice, also appear to be melting (FIG. 9-14). These factors could cause an additional rise in global sea levels, which would alter the shapes of the continents and sink low-lying barrier islands and atolls. For every foot that the sea level rises, 100 to 1,000 feet of shoreline disappears.

Fig. 9-13. Avalanche on Sherman Glacier, Cordova district, Alaska, caused by the August 24, 1964 Alaskan earthquake.

Sea level trends are estimated from tidal gage records at stations spread around the shorelines of the world. In some areas like Louisiana, the level of the sea has risen as much as 3 feet per century. Louisiana is losing about 6,000 acres of land each year to the encroaching sea. The beaches along North Carolina are retreating at a rate of 4 to 5 feet

Fig. 9-14. Meltwater from glaciers in Glacier National Park, Montana.

		TABLE 9-1. Major Changes in Sea Level	
DATE	SEA LEVEL	HISTORICAL EVENT	
2200 B.C.	low		
1600 B.C.	high	Coastal forest in Britain inundated by the sea.	
1400 B.C.	low		
1200 B.C.	high	Egyptian ruler Ramses II builds first Suez canal.	
500 B.C.	low	Many Greek and Phoenician ports built around this time are now under water.	
200 B.C.	normal		
A.D. 100	high	Port constructed well inland of present-day Haifa, Israel.	
A.D. 200	normal		
A.D. 400	high		
A.D. 600	low	Port of Ravenna, Italy, becomes land-locked. Venice is built and is presently being inundated by the Adriatic Sea.	
A.D. 800	high		
A.D. 1200	low	Europeans exploit low-lying salt marshes.	
A.D. 1400	high	Extensive flooding in low countries along the North Sea. The Dutch begin building dikes.	

per year. The higher sea levels are due in part to the land sinking as the increased weight of the water presses down on the continental shelf. In other regions like Scandinavia, the sea level has dropped as much as 3 feet per century because melting glaciers are reducing the weight on the land.

One of the first signs that rising global temperatures have started to warm the ocean was revealed in satellite measurements of the extent of the polar sea ice, which shrank by about 6 percent between 1973 and 1987. Sea ice forms a frozen band around Antarctica and covers most of the Arctic Ocean during the winter season in each hemisphere. The total surface area of the ice appears not to have changed significantly, but the extent to which the ice pack reaches outward from the poles has diminished. The ice obtains its maximum extent during spring in the Southern Hemisphere, when the Antarctic ice breaks up and the Arctic ice starts to spread. As the ocean continues to warm, the ice could melt closer to the poles and further reduce the perimeter of sea ice.

Most countries will feel the adverse effects of rising sea levels. If the present melting continues, the sea could rise 6 feet by the middle of the next century. Large tracks of coastal land would simply disappear along with shallow barrier islands and coral reefs. Low-lying fertile deltas that support millions of people would be drowned. Delicate wetlands, where many species of marine life hatch their young, would be reclaimed by the ocean. Vulnerable coastal cities would have to move farther inland or build protective walls against an angry sea.

THE NEXT ICE AGE

About 2.4 million years ago, when the surface waters of the ocean cooled dramatically, a series of ice ages began (TABLE 9-2). They occurred about every 100,000 years with an average of 90,000 years of glaciation and an interglacial period of 10,000 years. The ice age cycle responds to the Earth's orbital variations, which include a change in the shape of the orbit around the Sun from nearly a circle to an ellipse every 100,000 years, varying angles of the rotation axis from minimum to maximum tilt every 41,000 years, and the precession of the rotation axis in a complete circle every 21,000 years (FIG. 9-15).

TABLE 9-2. Chronology of the Major Ice Ages

TIME IN YEARS	EVENT
2 billion	First major ice age.
700 million	The great Precambrian ice age.
230 million	The great Permian ice age.
230 – 65 million	Interval of warm and relatively uniform climate.
65 million	Climate deteriorates, poles become much colder.
30 million	First major glacial episode in Antarctica.
15 million	Second major glacial episode in Antarctica.
4 million	Ice covers the Arctic Ocean.
2 million	First glacial episode in Northern Hemisphere.
1 million	First major interglacial.
100,000	Most recent glacial episode.
20,000 – 18,000	Last glacial maximum.
15,000 – 10,000	Melting of ice sheets.
10,000 – present	Present interglacial.

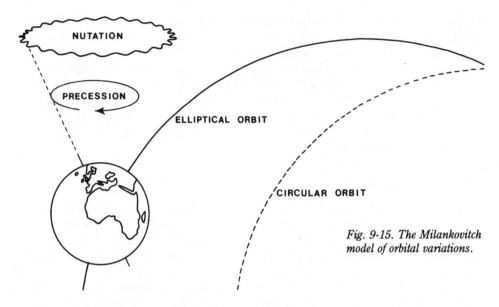

Fig. 9-15. The Milankovitch model of orbital variations.

The orbital motions do not alter the total amount of solar radiation that reaches the Earth during a year. They only alter the amount of solar energy that reaches different latitudes during different seasons. If the higher latitudes receive slightly less solar radiation during the summer months and slightly more radiation during the winter months, much of the winter's snowfall might fail to melt during the summer season. The addition of snow the following winter would then pile on top of the existing snow and bring on the full force of an ice age, possibly within as little as 200 years.

The last ice age began about 100,000 years ago and ended abruptly, in geological terms, about 10,000 years ago, when our present interglacial period began. The temperatures of this interglacial peaked around 6,000 years ago during the Climatic Optimum. Afterwards, temperatures dropped and the world became drier. The present deserts began to form about 4,000 years ago. Around 1,000 years ago, the world warmed up again during what is called the Medieval Climate Maximum. Five hundred years later, the world plunged into the Little Ice Age, which lasted for about 350 years. Then the world warmed again until 1938, cooled until 1976, and presently appears to be in another warming period.

These temperature fluctuations were mostly the result of natural climate variations and were outside the influence of man. The atmosphere responded to additional warming by expanding, which provided a larger area from which excess heat could be radiated out into space and further cooled the upper atmosphere.

According to the theory of orbital variations and geological evidence secured from sediments in ocean bottom cores and from ancient coral growth, the next ice age is overdue. The interglacial period prior to the last ice age had higher atmospheric concentrations of carbon dioxide and was warmer than the present interglacial. The warmer sea temperatures melted the polar ice caps and caused global sea levels to rise some 20 feet above their present level. Yet this did not prevent the onset of the last ice age and suggests that the present warming trend could actually trigger the onset of a new ice age.

10

Global Warming

RECENT droughts, floods, and unusual weather patterns have been blamed on the greenhouse effect. Atmospheric carbon dioxide levels have increased by about a third since the beginning of the industrial era. The global average temperature has risen over 1 degree Fahrenheit since the beginning of this century. By the middle of the next century, the amount of carbon dioxide in the atmosphere could double and global surface temperatures could rise as much as 5 to 10 degrees. This could dramatically change worldwide precipitation patterns, enlarge the world's deserts, and drastically reduce agricultural output. Any or all of which would be disastrous.

One-third of the carbon dioxide released into the atmosphere is from the destruction of tropical rain forests and the extension of agriculture. The rest comes from the combustion of fossil fuels, which produces an annual average of 1 ton of carbon for each of the world's 5 billion people. The consumption of fossil fuels, especially in the developing countries, is expected to continue to grow well into the next century with a concurrent increase in atmospheric carbon dioxide.

Increasing amounts of atmospheric carbon dioxide and other greenhouse gasses, principally methane, tend to warm the Earth, energize the atmosphere, and invigorate the hydrological cycle. The result could be more intense storm systems, and shifting precipitation patterns, which could cause serious drought and desertification in some areas and severe flooding in others. The rise in global temperatures could also melt the polar ice caps and raise sea levels enough to flood coastal regions several miles inland. A warmer Earth would, therefore, appear to be of no particular benefit to mankind.

THE CLIMATE PUZZLE

Extreme and often record-breaking weather events have recently occurred worldwide. There have been heat waves in American and central European cities, floods in Africa that interrupted nearly 2 decades of drought, and almost continuous rain and cold in the middle of summer at other places. The decade of the 1980s has had six of the hottest years of the century, even surpassing the Dust Bowl years of the 1930s. This might be a symptom of a global climate change that is caused by the chemical pollution of the atmosphere. The climate variability, however, is such that the strange weather can still be a reflection of natural variation. As yet, no sign of climate change has occurred that can be unquestionably blamed on the greenhouse effect.

There might be unknown moderating factors that can cancel out or at least lessen the greenhouse effect. Scientists have yet to determine where all the carbon dioxide being produced today is going. Only half the carbon dioxide produced by the burning of fossil fuels and deforestation has been found in the atmosphere or the ocean. Gaseous sulfur produced by marine single-celled plants called plankton might help counter human-induced global warming by partially regulating the Earth's temperature. The sulfur gas emissions could increase the concentration of cloud-forming particles, which could make clouds whiter and therefore more reflective. This in turn would lower global temperatures. Volcanic eruptions, decreasing solar activity, and decreasing stratospheric concentrations of ozone could induce some additional cooling.

It must also be kept in mind that the climate is always changing, even without man's contributions. During the last warm interglacial period about 125,000 years ago, the climate was much warmer than it is today. Sea levels were 20 feet higher due to the melting of the ice caps. The climate 10,000 years ago, at the beginning of the present interglacial, was significantly different from that of the previous 10,000 years, at the height of the last ice age. The climate of the last 200 years was much different than that of the previous 200 years during the Little Ice Age. The climate of the last 50 years was much warmer than that of the previous 50 years, when alpine glaciers reached their southern most extent. Usually, climate changes were slow enough in the past for the biological world to adapt. When climate changes were too abrupt, however, species became extinct.

The theory of climate change has still not developed enough to provide all the answers to the importance of the greenhouse effect. Few scientists publicly support the theory, but they also don't loudly deny it. Most have a wait and see attitude. More research is needed on atmospheric physics and air-sea interactions. Much needed information about the Earth will be collected by advanced satellite technology. The most powerful computers ever devised will be needed to model the data. When the data is analyzed, it could take perhaps a decade or more to sort it out.

If an upward trend in temperature continues well into the next decade, then scientists will be more certain that it is tied to the greenhouse effect. It is uncertain, however, whether the human race has that much time. If we wait too long to enact corrective measures, more drastic steps may be needed to counter global warming in the future.

Large-scale human intervention might be required to preserve plant and animal species that are threatened by global climate change, especially if the change happens too quickly. The two response strategies for combating climate change are adaptation, which might involve anything from moving to a cooler climate to building coastal defenses against a rising sea, and limitation, which directly involves limiting or reducing the emissions of greenhouse gasses. Perhaps a prudent response to climate change would utilize both of these measures.

Conservation can go a long way toward curtailing the effects of global warming. In large part, the consequences of conservation will be improved energy efficiency and the development of nonpolluting substitute energy sources. Conservation can only attack a portion of the carbon dioxide problem, however, not solve it. The world also needs a constitution for the atmosphere similar to the law of the sea because one nation's pollution inevitably affects all nations.

CARBON DIOXIDE

The concentration of carbon dioxide in the atmosphere has increased from 265 parts per million (ppm) in preindustrial times to 315 ppm in 1958, 335 ppm in 1978, 345 ppm in 1986, and 365 ppm in 1989. Sometime between 2020 and 2070, if present trends continue, the concentration of carbon dioxide in the atmosphere could be double the current value, which could increase the global mean surface temperature on average 5 degrees Fahrenheit, with some areas experiencing as much as a 10 degree increase (FIG. 10-1).

The carbon dioxide content of the atmosphere fluctuates with the seasons (FIG. 10-2). It peaks in late winter and falls to a minimum level at the end of summer. This is because plants draw carbon dioxide out of the atmosphere during the growing season and store it in their tissues in the form of carbohydrates. The world's great forests have a pronounced influence on the content of atmospheric carbon dioxide. Much of the seasonal variation in the atmospheric concentration correlates with the rapid rise of photosynthesis during the summer. Forests conduct more photosynthesis worldwide than any other form of vegetation. Forests store enough carbon to substantially affect the carbon dioxide content of the atmosphere.

Although the mechanisms of the greenhouse effect are not fully understood, the results of a steady increase in atmospheric carbon dioxide would probably be catastrophic if other moderating factors did not come into play. An increase in the average world temperature could enlarge the desert and semidesert regions and significantly affect agriculture. On the other hand, an increase in atmospheric carbon dioxide, which acts as a sort of fertilizer, could also encourage the growth of plants and cause a greening of the Earth. Which direction the climate will follow still remains a riddle.

The long-term increase in atmospheric carbon dioxide, as much as 25 percent since 1860, is the result of an accelerated release of carbon dioxide by the combustion of fossil fuels. Americans release nearly 6 tons per person per year. This amounts to 1.2 billion tons, or about one-quarter of the world's total output. The atmosphere at present holds about 700 billion tons of carbon. Humans alone are increasing the amount of atmospheric carbon by nearly 1 percent annually.

Some of the carbon is taken out of the atmosphere by biological, hydrological, and geological processes. This reduces the average annual increase of atmospheric carbon dioxide from man's activity to about one-half of 1 percent or 3 million tons. The biota on the surface and humus in the soil hold 40 times as much carbon as the entire atmosphere. The harvest of forests, the extension of agriculture, and the destruction of wetlands speed the decay of humus, which is transformed into carbon dioxide and released

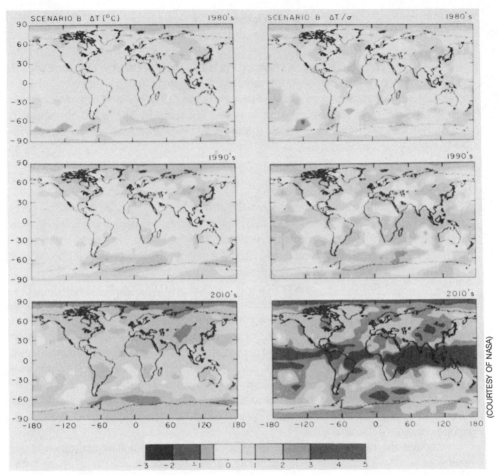

Fig. 10-1. Mean global temperature change for the decades of the 1980s, 1990s, and 2010s.

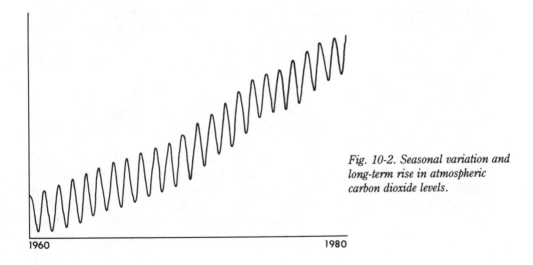

Fig. 10-2. Seasonal variation and long-term rise in atmospheric carbon dioxide levels.

into the atmosphere. Agricultural lands do not store as much carbon as the forests they replace. The soil does, however, release large quantities of carbon dioxide when it is cultivated. Each time the soil is turned over, new organic matter is exposed to the atmosphere.

The largest store of carbon dioxide is held by the ocean, which contains as much as 60 times more carbon than the atmosphere. Carbon dioxide enters the ocean by surface wave action. The concentration of carbon dioxide in the mixed layer of the ocean, the top 250 feet, is as much as the entire atmosphere itself. In this region of the ocean, microorganisms use the carbon dioxide in the form of bicarbonite to make their skeletons and shells. When the animal dies, its hard shell or skeleton settles on the bottom of the ocean where it contributes to the formation of carbonate rock. If the calcium carbonate falls to greater depths, it is dissolved in the cold deep waters of the abyssal. The abyssal region, by virtue of its great volume, holds the vast majority of free carbon dioxide. Due to the upwelling of carbon dioxide-rich waters from the deep ocean, carbon dioxide concentrations are greater at the equator than at other latitudes.

The capacity of the abyssal to absorb carbon dioxide is almost limitless. Nevertheless, carbon dioxide moves from the atmosphere, through the mixed layer of the ocean and into the oceanic depths very slowly and at a nearly constant rate. Unfortunately, the rate of absorption appears to be only about half the rate of release from the combustion of fossil fuels. The problem is worse because the biota is also a net source of atmospheric carbon dioxide. It releases the gas in amounts equal to the rate generated by the combustion of fossil fuels. Without man's contribution, the atmosphere and the ocean would be in equilibrium. Humans are short-circuiting the carbon cycle by placing the atmosphere and ocean out of balance.

An increase in surface temperature could have a worldwide effect on precipitation (FIG. 10-3). Areas between 20 and 50 degrees north latitude and 10 and 30 degrees south latitude would experience a marked decrease in precipitation which would encourage the spread of deserts. Presently, the United States has more than half a billion acres of arid and semiarid land. There are even larger areas in Africa, Australia, and South America. Changes in precipitation patterns would have profound effects on the distribution of water resources in agricultural areas that depend on irrigation.

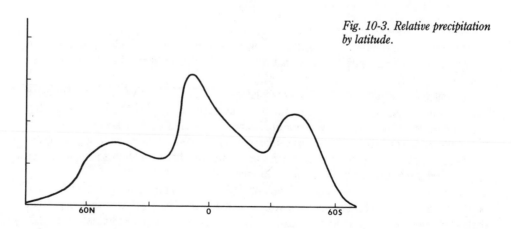

Fig. 10-3. Relative precipitation by latitude.

60N   0   60S

Not only would rainfall diminish, but higher temperatures would augment evaporation. The flow of rivers could decline by 50 percent or more with some rivers drying out altogether. Major groundwater supplies would also be adversely affected. Deep water wells needed for irrigation could go dry. Other areas would receive a large increase in precipitation, which could cause extensive flooding to prime agricultural lands (FIG. 10-4).

(COURTESY OF USDA-SOIL CONSERVATION SERVICE)

Fig. 10-4. The August 1955 Connecticut flood that submerged much of the farmland.

VIOLENT STORMS

An increase in temperatures could energize the atmosphere and add extra power to storm systems. Hardly a day goes by that there is not a major storm that takes lives and destroys property in some part of the world.

Hurricanes are nature's most spectacular as well as destructive storms. They are more common in summer and autumn months when the sun can heat the sea well to the north or south of the equator. When a hurricane makes landfall, it is accompanied by a tremendous storm surge or tidal wave that wrecks property and erodes beach fronts. Winds of 100 miles per hour or more push water onto the shore. Low pressure in the eye of the hurricane pulls water up into a mound several feet high.

Tornadoes strike sporadically and violently (FIG. 10-5). They generate the strongest of all surface winds and cause more deaths annually in the United States than any other natural phenomenon. Tornadoes develop in the spring and to a lesser extent in the fall.

Fig. 10-5. Tornado damage at Birmingham, Alabama, on April 4, 1975.

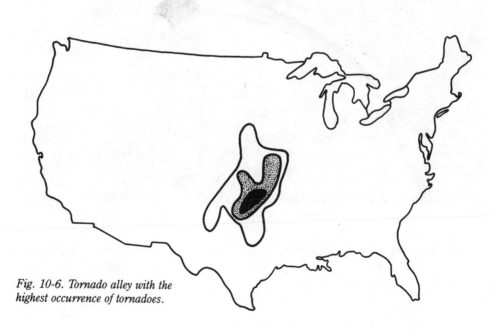

Fig. 10-6. Tornado alley with the highest occurrence of tornadoes.

The United States is the world's tornado hot spot with about 700 tornadoes yearly. The central and southeastern portions of the country, known as tornado alley (FIG. 10-6), are particularly hard hit. Australia comes in a close second.

It is estimated that at any given moment, there are nearly 2,000 thunderstorms in progress around the world. Their sheer numbers make thunderstorms the primary balancers of the Earth's heat budget. They are most common in spring and summer months, although they do occur infrequently in winter. On average, 100 Americans are killed and 250 are seriously injured by thunderstorms every year. Property losses are estimated in the hundreds of millions of dollars annually. Thunderstorms, which are generated by temperature imbalances in the atmosphere, are violent examples of the upward transfer of heat flow. They can be seriously affected by greenhouse warming.

The frequency of dust storms is expected to escalate as the ground becomes hotter and drier and winds become more blustery. These storms present a solid wall of dust (FIG. 10-7) that blows at speeds upwards of 60 miles per hour. The dust can reach several thousand feet in altitude and stretch for hundreds of miles. The land is scoured by the winds. Several inches of soil can be lifted to other regions, even across the ocean. Giant dust storms arise in the deserts of Africa, Arabia, central Asia, Australia, and the Americas. The most obvious threat of dust storms is soil erosion.

(COURTESY OF USDA-SOIL CONSERVATION SERVICE)

Fig. 10-7. A massive dust storm in the western plains of the U.S.

Of all of nature's violence, nothing can compare to lightning for its instantaneous release of intense energy (FIG. 10-8). Lightning is very destructive to structures, causes most forest fires, and kills more people than any other weather phenomenon. Over the

Fig. 10-8. Tremendous lightning bolts are the major cause of forest fires in the U.S.

past 20 years, an average of 100 people a year have been killed by lightning in the United States. As the atmosphere becomes more turbulent because of greenhouse warming, the number of lightning strikes is expected to increase. This adds an additional danger of forest fires.

Along with an increase in the frequency of storms from greenhouse warming, there might also be an increase in floods. There are over 3.5 million miles of streams in North America. About 6 percent of the land that adjoins these streams is prone to flooding (FIG. 10-9). Flash floods are among the most severe. They occur when a violent thunderstorm dumps a large quantity of rain on a relatively small area in a very short time. They are particularly common in the mountainous areas and desert regions of the western United States. Torrential downpours and tidal floods that accompany hurricanes cause more damage and take more lives than other forms of flooding.

One of the most serious effects of global warming could be the melting of the ice caps. The release of meltwater raises sea levels and floods coastal regions. To combat this effect, seawalls have been constructed against the vicissitudes of the ocean. Steep waves that accompany sea storms seriously erode sand dunes and sea cliffs. The constant pounding of the surf also erodes most man-made defenses (FIG. 10-10). Upwards of 90 percent of America's once-sandy beaches are sinking beneath the waves. Barrier islands and sandbars that run along the East Coast and along the eastern Texas coast are disappearing at alarming rates. Sea cliffs are eroding back several feet a year. In California, huge chunks of land are falling into the sea. Most attempts at stopping beach erosion are usually defeated by nature.

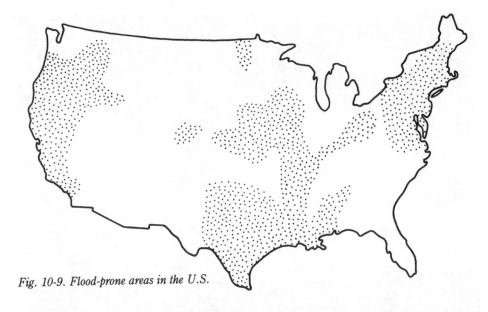

Fig. 10-9. Flood-prone areas in the U.S.

(PHOTO BY P.W. KOCH, COURTESY OF USDA-SOIL CONSERVATION SERVICE)

Fig. 10-10. Efforts to halt beach erosion include a system of groins to trap the sand.

THE CHANGING CLIMATE

There seems to be little doubt that greenhouse warming has a substantial effect on the climate. Without the quantities of greenhouse gasses in the atmosphere, surface temperatures would fall about 60 degrees Fahrenheit and the planet would be covered with ice.

Scientists have known about the mechanics of the greenhouse effect since before the turn of this century. The Swedish chemist Svante Arrhenius predicted the effects of atmospheric carbon dioxide on the climate in 1896. He concluded that past glacial epochs might have occurred largely because of a reduction in atmospheric carbon dioxide. Arrhenius also estimated that a doubling of the concentration of carbon dioxide in the atmosphere would cause a global warming of about 9 degrees Fahrenheit.

Within 50 to 100 years, the world could be hotter than it has been in the past million years. The warming will be greatest at the higher latitudes of the Northern Hemisphere. The greatest temperature increases will occur in the winter months. Evaporation rates will increase, which will change circulation patterns and dramatically affect the weather. Vast areas of once productive cropland could lose topsoil and become man-made deserts. At the present rate of destruction, many rain forests will be replaced by deserts.

What is unique about the current warming trend, which amounts to an increase of over 1 degree Fahrenheit this century, is its unprecedented speed. The present warming is 10 to 40 times faster than the average rate of warming that followed the last ice age. By the end of the Pleistocene epoch, between 14,000 and 10,000 years ago, the Earth warmed perhaps 5 to 10 degrees Fahrenheit. Although this temperature increase is similar to the predicted increase for the greenhouse effect, the major difference is that it was spread over a period of several thousand years and not compressed into less than a century (FIG. 10-11).

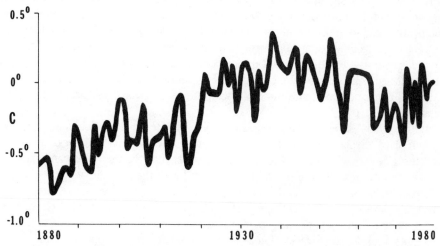

Fig. 10-11. One hundred years of temperature variations in the Northern Hemisphere.

TABLE 10-1. The Warmest, Wettest, and Windiest Cities in the U.S.

EXTREME	LOCATION	AVERAGE ANNUAL VALUE
Warmest	Key West, Florida	Mean temperature 78 deg. F
Coldest	International Falls, Minnesota	Mean temperature 36 deg. F
Sunniest	Yuma, Arizona	348 sunny days
Driest	Yuma, Arizona	2.7 inches of rainfall
Wettest	Quillayute, Washington	105 inches of rainfall
Rainiest	Quillayute, Washington	212 rainy days
Cloudiest	Quillayute, Washington	242 cloudy days
Snowiest	Blue Canyon, California	243 inches of snowfall
Windiest	Blue Hill, Massachusetts	Mean wind speed 15 mph

If present warming trends continue, by the end of the next century global temperatures could be as warm as they were during the Cretaceous period 100 million years ago. This was the hottest period in geological history. The continents at that time were bunched near the equator, however, so the climates of the two periods would not be the same.

Some areas, particularly in the Northern Hemisphere, will dry out because of the higher temperatures. This will increase the potential for massive forest fires. During the relatively warm and dry fifteenth and sixteenth centuries, major forest fires in North America appear to have occurred about once every nine years. Then, over the next three centuries, during the cooling period of the Little Ice Age, massive forest fires were less frequent, less intense, and occurred only about every 14 years. This has dire implications for us today. If greenhouse warming continues, major forest fires such as those that devastated half of Yellowstone National Park during the summer and fall of 1988 (FIG. 10-12) might become more frequent, with the resultant loss of forests and wildlife habitats.

At the other extreme, the southern tropics will face severe flooding, which will erode cropland, displace people, and generally cause an ecological disaster of unheard of proportions. Rivers will be forced to carry more water than their courses can handle and will take massive quantities of much needed topsoil with them to the sea.

Coastal regions, where half the human population lives, will feel the adverse effects of rising sea levels when the ice caps melt under increasing ocean temperatures. If the present melting continues, the sea could rise as much as 6 feet by the middle of the next century. As a result, large tracks of coastal land would disappear along with shallow barrier islands and coral reefs. Low-lying fertile deltas that support millions of people would vanish. Delicate wetlands, where many species of marine life hatch their young, would be reclaimed by the sea. Vulnerable coastal cities would have to move farther inland or build protective walls.

Fig. 10-12. Burned trees from the 1988 forest fire that devastated Yellowstone National Park.

The effects of global warming could last for centuries. Forests would creep northward. Other wildlife habitats, including arctic tundra, would disappear entirely. Many species will be unable to keep pace with the rapid climate changes. Those that are able to migrate could find their routes blocked by natural and man-made barriers. The warming would rearrange entire biological communities and cause many species to become extinct. Other species, commonly called pests, could overrun the land. These changes would diminish the diversity of the world's species and have an adverse affect on humans as the biosphere loses its ability to sustain life. Because of man's disruptive activities, the precious planet called Earth could become a totally alien world.

Glossary

abrasion—Erosion by friction, generally caused by rock particles carried by running water, ice, and wind.

absolute zero—The temperature at which all molecular motion ceases: −273 degrees Celsius or 0 degrees Kelvin.

absorption—The process by which radiant energy incident on any substance is retained and converted into heat or another form of energy.

abyssal—The deep ocean, generally over a mile in depth.

adiabatic—A change in temperature that occurs within an air mass as a result of pressure changes within the air mass that cause it to expand or contract without gain or loss of heat.

adsorption—The adhesion of a thin film or liquid or gas to the surface of a solid substance.

advection—The horizontal movement of air, moisture, or heat.

aerosol—A mass made of solid or liquid particles dispersed in air.

air mass—An extensive body of air whose horizontal distribution of temperature and moisture is nearly uniform.

airstream—A substantial body of air with the same characteristics flowing with general circulation.

albedo effect—The amount of sunlight reflected from an object.

angular momentum—The measure of an object or orbiting body to continue spinning.

anticyclone—The circulation of air around a central area of high pressure that is usually associated with settled weather; pressure rises steadily when an anticyclone is developing and falls when it is declining.

aphelion—The point at which the orbit of a planet is at its farthest point from the Sun. In the case of the Earth, it occurs in early July.

aquifer—A subterranian bed of sediments through which groundwater flows.

asteroid—A rocky or metallic body that orbits the Sun between Mars and Jupiter.

asthenosphere—A layer of the upper mantle that lies between 50 and 200 miles below the surface. It is more plastic than the rock above and below and might be in convective motion.

atmospheric pressure—The weight per unit area of the total mass of air above a given point; also called barometric pressure.

aurora—Luminous bands of colored light seen near the poles due to cosmic-ray bombardment of the upper atmosphere.

basalt—A volcanic rock that is dark in color and usually quite fluid in the molten state.

black body radiation—The electromagnetic radiation emitted by an ideal black body. It is the theoretical maximum amount of radiant energy at all wave lengths that can be emitted by a body at a given temperature.

blocking high—Any high-pressure center that remains stationary and effectively blocks the usual eastward progression of weather systems in the middle latitudes for several days.

blowout—A hollow caused by wind erosion.

carbonate—A mineral containing calcium carbonate such as limestone and dolostone.

Cenozoic era—An age known as recent life, which spans from 65 million years ago to the present.

center of mass—The point at which all the mass of a body or system of bodies might be considered for calculating the gravitational effect when a force is applied.

circulation—The flow pattern of moving air or water.

circum-Pacific belt—Active seismic regions on the rim of the Pacific plate that coincide with the ring of fire.

coalescence—The merging of two or more colliding water droplets into a single, larger drop.

cold front—The interface or transition zone between advancing cold air and retreating warm air.

condensation—The process whereby a substance changes from the vapor phase to liquid or solid phase; the opposite of evaporation.

conduction—The transmission of heat by direct contact through a material substance, as distinguished from convection, advection, and radiation.

continent—A slab of light, granitic rock that floats on the denser rocks of the upper mantle.

continental shelf—The offshore area of a continent in shallow sea.

continental slope—The transition from the continental margin to the deep sea basin.

convection—A circular, vertical flow of a fluid medium due to heating from below. As materials are heated they become less dense and rise while cooler, heavier materials sink.

convergence—A distribution of wind movement that results in a net inflow of air into a region such as a low-pressure area.

coral—Any of a large group of shallow-water, bottom-dwelling marine invertebrates, whose reef-building colonies are common in the warm waters of the tropics.

core—The central part of the Earth that has a radius of 2,300 miles and consists of a crystalline inner core and molten outer core composed of iron and nickel.

Coriolis effect—The apparent force that deflects the wind or a moving object and causes it to curve in relation to the rotation of the Earth.

cosmic rays—High-energy charged particles that enter the Earth's atmosphere from outer space.

Cretaceous period—A geologic period encompassing from 135 to 65 million years ago and commensurate with the age of the dinosaurs.

crust—The outer layers of the Earth's rocks.

currie point—The temperature at which the magnetic fields of iron atoms align to the Earth's magnetic field.

cyclone—The circulation of air around a central area of low pressure that is usually associated with unsettled weather. In tropical latitudes it can refer to an intense storm that does not attain full hurricane status.

density—The amount of any quantity per unit volume.

dew point—The temperature to which air, at a constant pressure and moisture content, must be cooled for saturation to occur.

diapir—The buoyant rise of a molten rock through heavier rock.

differentiation—The separation of solids or liquids according to their weight.

diffusion—The exchange of fluid substance and its properties between different regions in the fluid, as the result of small, almost random motions of the fluid.

drought—A period of abnormally dry weather sufficiently prolonged for the lack of water to cause serious deleterious effects on agricultural and other biological activities.

dry adiabatic lapse rate—The rate at which dry air cools with height when it is forced to rise into regions of lower pressure: 5.4 degrees Fahrenheit per 1,000 feet (1 degree Celsius per 100 meters). Dry sinking air warms at the same rate.

earthquake—The sudden breaking of crustal rocks.

ecliptic—The plane in which the Earth's orbit traces an elliptical path around the Sun.

electromagnetic radiation—The energy from the Sun that travels through the vacuum of space and reaches the Earth as electromagnetic waves.

electron—A negative particle of small mass orbiting the nucleus and equal in number to the proton.

element—A material consisting of only one type of atom.

eolian—A deposit of wind-blown sediment.

equinox—Either of the two points of intersection of the Sun's path and the plane of the Earth's equator.

estuary—A tidal inlet along a coast.

evaporation—The transformation of a liquid into a gas.

exosphere—The outermost portion of the atmosphere. Its lower boundary is at a height of 300 miles.

extrusive—Igneous volcanic rock that is ejected on the Earth's surface.

evolution—The tendency of physical and biological factors to change with time.

fault—The breaking of rocks caused by crustal motion.

fossil—Any remains, impression, or trace in rock of a plant or animal of a previous geological age.

frequency—The rate at which crests of any wave pass a given point.

Gondwana—A southern supercontinent of Paleozoic time, which consisted of Africa, South America, India, Australia, and Antarctica. It broke up into the present continents during the Mesozoic era.

granite—A coarse-grained, silica-rich rock that consists primarily of quartz and feldspars. It is the principal constituent of the continents and is believed to be derived from a molten state beneath the Earth's surface.

greenhouse effect—The global heating effect that is caused when the atmosphere is more transparent to incoming short-wave solar radiation than it is to outgoing long-wave radiation.

groundwater—The water derived from the atmosphere that percolates and circulates below the surface of the Earth.

Hadley cell—Atmospheric circulation that distributes air from the tropics to the poles. It is sustained by large-scale convection currents in which hot air is replaced by cooler air.

heat flow—The rate that heat energy flows from hot towards cold at a rate or flux equal to the temperature gradient times and the conductivity of the material in between.

helium—The second lightest and second most abundant element in the universe. It is composed of two protons and two neutrons.

high—An area of high atmospheric pressure within a closed circulation system—an anticyclone.

hydrocarbon—A molecule that consists of carbon chains with attached hydrogen atoms.

hydrogen—Lightest and most abundant element in the universe. It is composed of one proton and one electron.

igneous rocks—All rocks that have solidified from a molten state.

inertia—Inherent resistance to applied force.

infrared—Invisible light with a wavelength between red light and radio waves.

insolation—Solar radiation impinging on the Earth. The word is a contraction of incoming solar radiation.

interglacial—A warming period between glacial periods.

intertropical convergence zone—The axis along which the northeast trade winds of the Northern Hemisphere meet the southeast trade winds of the Southern Hemisphere.

inversion—A departure from the usual decrease or increase with altitude of the value of an atmospheric property.

ionization—The process whereby electrons are torn off previously neutral atoms.

ionosphere—The atmospheric shell that is characterized by high ion density. It extends from about 40 miles to very high regions of the atmosphere.

iridium—A rare isotope of platinum, relatively abundant on meteorites.

island arc—The volcanoes landward of a subduction zone, parallel to the trench, and above the melting zone of a subduction zone.

isotope—A variety of an element that has a different number of neutrons in the nucleus, but exhibits identical chemical behavior.

jet stream—Relatively strong winds concentrated within a narrow belt that is usually found in the tropopause.

Kelvin—A temperature scale that is similar to the Centigrade scale. Its zero point is placed at absolute zero, or −273 degrees C.

kinetic energy—The energy that a moving body possesses as a consequence of its motion.

landslide—Rapid downhill movement of Earth materials often triggered by earthquakes.

lapse rate—The decrease of an atmospheric variable, usually temperature, with height.

latent heat—Heat absorbed when a solid changes to a liquid or a liquid to a gas with no change in temperature, or the heat released in the reverse transformations.

Laurasia—The northern supercontinent of the Paleozoic era that consisted of North America, Europe, and Asia.

lava—Molten magma after it has flowed out onto the surface.

light-year—The distance that electromagnetic radiation, principally light waves, can travel in a vacuum in one year or approximately 6 trillion miles.

limestone—A sedimentary rock composed of calcium carbonate that is secreted from seawater by invertebrates, whose skeletons compose the bulk of deposits.

lithosphere—A rigid outer layer of the mantle, typically about 60 miles thick. It is over ridden by the continental and oceanic crusts and is divided into segments called plates.

low—An area of low atmospheric pressure—a cyclone or a depression.

lysocline—The ocean depth below which the rate of dissolution just exceeds the rate of deposition of the dead shells of calcareous organisms.

magma—A molten rock material generated within the Earth. It is the constituent of igneous rocks, including volcanic eruptions.

magnetic field reversal—A reversal of the polarity of the Earth's magnetic poles. This has occurred intermittantly throughout geological time.

magnetometer—A device used to measure the intensity and direction of the magnetic field.

magnetosphere—The region of the Earth's upper atmosphere in which the Earth's magnetic field controls the motion of ionized particles.

mantle—The part of the Earth below the crust and above the core that is composed of dense iron-magnesium rich rocks.

mean temperature—The average of any series of temperatures observed over a period of time.

mesosphere—A region of the Earth's atmosphere between the stratosphere and thermosphere, which extends 24 to 48 miles above the Earth's surface. Also, the rigid part of the Earth's mantle below the asthenosphere and above the core.

Mesozoic era—Literally the period of middle life. It refers to the period between 230 and 65 million years ago.

metamorphic rock—A rock crystallized from previous igneous, metamorphic, or sedimentary rocks created under conditions of intense temperatures and pressures without melting.

meteorite—A metallic or stony body from space that enters the Earth's atmosphere and impacts on the Earth's surface.

methane—A hydrocarbon gas liberated by the decomposition of organic matter.

micron—A unit of measurement equivalent to one-thousandth of a millimeter.

midocean ridge—A submarine ridge along a divergent plate boundary where a new ocean floor is created by the upwelling of mantle material.

monsoon—A seasonal wind accompanying temperature changes over land and water from one season of the year to another.

moraine—A ridge of erosional debris deposited by the melting margin of a glacier.

nebula—An extended astronomical object with a cloud-like appearance. Some nebulae are galaxies; others are clouds of dust and gas within our galaxy.

neutrino—A small electrically neutral particle that has weak nuclear and gravitational interactions.

neutron—A particle that has no electrical charge and roughly the same weight as the positively charged proton, both of which are found in the nucleus of an atom.

Oort cloud—The collection of comets that orbits close to the Earth once every 26 million years and showers the planet with meteorites, possibly causing extinctions.

orogeny—An episode of mountain building.

outgassing—The loss of gas within a planet as opposed to degassing, or the loss of gas from meteorites.

ozone—A molecule consisting of three atoms of oxygen that exists in the upper atmosphere above the tropopause and filters out ultraviolet radiation from the Sun.

paleomagnetism—The study of the Earth's magnetic field, including the position and polarity of the poles in the past.

paleontology—The study of ancient life forms, based on the fossil record of plants and animals.

Paleozoic era—The period of ancient life that existed between 570 and 230 million years ago.

Pangaea—An ancient supercontinent that included all the landmass of the Earth.

Panthalassa—The great world ocean that surrounded Pangaea.

perihelion—The point at which the orbit of a planet is at its nearest to the Sun. In the case of the Earth, it occurs in early January.

permafrost—Permanently frozen ground.

permeability—The ability to transfer fluid through cracks, pores, and interconnected spaces within a rock.

phenology—The study of the times of recurring natural phenomena in relation to climatic conditions.

photon—A packet of electromagnetic energy, generally viewed as a particle.

photosynthesis—The process by which plants create carbohydrates from carbon dioxide, water, and sunlight.

plate tectonics—The theory that accounts for the major features of the Earth's surface in terms of the interaction of lithospheric plates.

polar air—An air mass conditioned over the tundra or snow-covered terrain of high latitudes.

polar front—A semipermanent discontinuity that separates the cold polar easterly winds and relatively warm westerly winds of the middle latitudes.

porosity—The percent of a rock that consists of pore spaces between crystals and grains, usually filled with water.

precession—The slow change in direction of the Earth's axis of rotation due to gravitational action of the Moon on the Earth.

precipitation—Products of condensation that fall from clouds as rain, snow, hail, or drizzle.

proton—A particle with a positive charge in the nucleus of an atom.

radiation—The process by which energy from the Sun is propagated through a vacuum of space as electromagnetic waves. A method, along with conduction and convection, of transporting heat.

radioactivity—An atomic reaction that releases detectable radioactive particles.

radiometric dating—The determination of how long an object has existed by chemical analysis of stable versus unstable radioactive elements.

regression—A fall in sea level that exposes continental shelves to erosion.

relative humidity—The ratio of the amount of moisture in the air to the amount that the air would hold at the same temperature and pressure if it were saturated. It is usually expressed as a percentage.

rift valley—The center of an extensional spreading center where continental or oceanic plate separation is occurring.

sandstone—A sedimentary rock that consists of cemented sand grains.

saturated air—Air that contains the maximum amount of water vapor it can hold at a given pressure and temperature. Saturated air has a relative humidity of 100 percent.

seafloor spreading—The theory that the ocean floor is created by the separation of lithospheric plates along the **midocean ridges**, with new oceanic crust forming from mantle material that rises from the mantle to fill the rift.

seamount—An undersea volcano that never reached the surface of the ocean and so does not have a flat erosional top.

seismic sea wave—An ocean wave related to an undersea earthquake.

shield—Areas of the exposed Precambrian nucleus of a continent.

solar flare—A short-lived bright event on the Sun's surface that causes greater ionization of the Earth's upper atmosphere due to an increase in ultraviolet light.

solar wind—An outflow of particles from the Sun that represents the expansion of the corona.

solstices—The occurrence twice yearly when the apparent distance of the Sun from the Earth's equator is at its greatest. During summer solstice, the Sun appears to be at its most northerly position on June 22. The Sun is then directly overhead at a latitude of 23.5 degrees North. During winter solstice, the Sun appears to be at its most southerly position on December 22. The Sun is then directly overhead at a latitude of 23.5 degrees South.

soluble—Refers to a substance that dissolves in water.

storm surge—An abnormal rise of the water level along a shore as a result of wind flow in a storm.

stratosphere—An upper layer of the atmosphere above the troposphere that is between 12 to 30 miles above the Earth's surface. The air in this layer is usually stable and the temperature increases with height.

subduction zone—An area where the oceanic plate dives below a continental plate into the asthenosphere. Ocean trenches are the surface expression of a subduction zone.

sublimation—A process by which a gas is changed into a solid or a solid is changed to a gas without going through the liquid state.

subsidence—The descending motion aloft of a body of air, usually within an anticyclone. It causes the lower layers of the atmosphere to spread out and warm.

sunspot—A region on the Sun's surface that is cooler than surrounding regions. It can effect radio transmissions on Earth.

supercooling—The cooling of a liquid below its freezing point without it becoming a solid.

supernova—An enormous stellar explosion in which all but the inner core of a star is blown off into interstellar space. It produces as much energy in a few days as the Sun does in a billion years.

synod—The alignment of the Sun, planets, and their accompanying moons.

tectonic activity—The formation of the Earth's crust by large scale landmass movements throughout geological time.

temperature inversion—A layer of the atmosphere in which the temperature increases with altitude as opposed to the normal tendency for temperature to decrease with altitude.

tephra—All clastic material from dust particles to large chunks that are expelled from volcanoes during eruptions.

Tethys Sea—The hypothetical mid-latitude area of the oceans separating the northern and southern continents of Gondwanaland and Laurasia some hundreds of million years ago.

thermalsphere—The outermost layer of the atmosphere in which the temperature increases regularly with height.

tide—A bulge in the ocean that is produced by the Moon's gravitational forces on the Earth's oceans. The rotation of the Earth beneath this bulge causes the sea level to rise and fall.

transform fault—A major plate boundary that is formed when two plates move across each other along a fault.

transgression—A rise in sea level that causes flooding of the shallow edges of continental margins.

trench—A topographic feature that is formed when the seafloor plunges into the mantle along a line of subduction.

tropical cyclone—A low-pressure area that originates in the tropics, has a warm central core, and often develops an eye.

troposphere—The lowest 6 to 12 miles of the Earth's atmosphere that is characterized by decreasing temperature with height.

typhoon—Severe tropical storms in the western Pacific Ocean similar in structure to a hurricane.

ultraviolet—The invisible light that has a wavelength shorter than visible light and longer than x-rays.

uniformitarianism—The belief that the slow processes that shape the Earth's surface have acted essentially unchanged throughout geological time.

Van Allen belts—Regions of high-energy particles that are trapped by the Earth's magnetic field.

viscosity—The resistance of a liquid to flow.

volcano—A fissure or vent in the crust through which molten rock rises to the surface to form a mountain.

warm front—The boundary of an advancing current of relatively warm air that is displacing a retreating colder air mass.

water vapor—Atmospheric moisture in the invisible gaseous phase.

Bibliography

THE PRIMORDIAL GREENHOUSE

Holland, H. D., B. Lazar, and M. McCaffrey. "Evolution of the atmosphere and oceans." *Nature* Vol. 320 (March 6, 1986): 27–33.

Kerr, Richard A. "How to Make a Warm Cretaceous Climate." *Science* Vol. 223 (February 17, 1984): 677–678.

Officer, Charles B., and Charles L. Drake. "The Cretaceous-Tertiary Transition." *Science* Vol. 219 (March 25, 1983): 1383–1390.

Rossow, William B., Ann Henderson-Sellers, and Stephen K. Weinreich. "Cloud Feedback: A Stabilizing Effect for the Early Earth?" *Science* Vol. 217 (September 24, 1982): 1245–1247.

Toon, Owen B. and Steve Olson. "The Warm Earth." *Science 85* (October 1985): 50–57.

Towe, Kenneth M. "Earth's Early Atmosphere." *Science* Vol. 235 (January 23, 1987): 415.

Weisburd, Stefi. "Forests made the world frigid?" *Science News* Vol. 131 (January 3, 1987): 9.

THE ROCK CYCLE

Berner, Robert A. and Antonio C. Lasaga. "Modeling the Geochemical Carbon Cycle." *Scientific American* Vol. 260 (March 1989): 74–81.

Bonatti, Enrico. "The Rifting of Continents." *Scientific American* Vol. 256 (March 1987): 97–103.

Broecker, Wallace S. "Carbon dioxide circulation through ocean and atmosphere." *Nature* Vol. 308 (April 12, 1984): 602.

Heppenheimer, T. A. "Journey to the Center of the Earth." *Discover* Vol. 8 (November 1987): 86–93.

Kerr, Richard A. "No Longer Willful, Gaia Becomes Respectful." *Science* Vol. 240 (April 22, 1988): 393–395.

_____. "The Carbon Cycle and Climate Warming." *Science* Vol. 222 (December 9, 1983): 1107–1108.

Monastersky, Richard. "The Whole-Earth Syndrome." *Science News* Vol. 133 (June 11, 1988): 378–380).

Nance, R. Damian, Thomas R. Worsley, and Judith B. Moody. "The Supercontinent Cycle." *Scientific American* Vol. 259 (July 1988): 72–79

White, Robert S. and Dan P. McKenzie. "Volcanism at Rifts." *Scientific American* Vol. 261 (July 1989): 62–71.

THE WATER CYCLE

Abelson, Philip H. "Climate and Water." *Science* Vol. 243 (January 27, 1989): 461.

Ambroggi, Robert P. "Water." *Scientific American* Vol. 243 (September 1980): 101–115.

Borecker, Wallace S. "The Ocean." *Scientific American* Vol. 249 (September 1983): 146–160.

Ellsaesser, Hugh W. "The Greenhouse Effect: Science Fiction?" *Consumer's Research* Vol. 71 (November 1988): 27–31.

Hollister, Charles D., Arthur R. M. Nowell, and Peter A. Jumars. "The Dynamic Abyss." *Scientific American* Vol. 250 (March 1984): 42–53.

Kerr, Richard A. "Linking Earth, Ocean, and Air at the AGU." *Science* Vol. 239 (January 15, 1988): 259–260.

_____. "Ocean Crust's Role in Making Seawater." *Science* Vol. 239 (January 15, 1988): 260.

Monastersky, Richard. "Recent ocean warming: Are satellites right?" *Science News* Vol. 135 (April 22, 1989): 247.

THE WEATHER MAKER

Graham, N. E. and T. P. Barnett. "Sea Surface Temperature, Surface Wind Divergence, and Convection over Tropical Oceans." *Science* Vol. 238 (October 30, 1987): 657–659.

Kerr, Richard A. "How to Fix the Clouds in Greenhouse Models." *Science* Vol. 243 (January 6, 1989): 28–29.

_____. "Sunspot-Weather Correlation Found." *Science* Vol. 238 (October 23, 1987): 479–480.

Lo Presto, James Charles. "Looking Inside the Sun." *Astronomy* Vol. 17 (March 1989): 22–30.

Monastersky, Richard. "Clouds clearing from climate predictions." *Science News* Vol. 135 (January 7, 1989): 6.

Ramage, Colin S. "El Niño." *Scientific American* Vol. 254 (June 1986): 77–83.

Webster, Peter J. "Monsoons." *Scientific American* Vol. 245 (August 1981): 109–118.

FOSSIL FUEL COMBUSTION

Ableson, Philip H. "Future Supplies of Energy and Minerals." *Science* Vol. 231 (February 14, 1986): 657.

Flower, Andrew R. "World Oil Production." *Scientific American* Vol. 238 (March 1978): 42–48.

Griffith, Edward D. and Alan W. Clarke. "World Coal Production." *Scientific American* Vol. 240 (January 1979): 38–47.

Hapgood, Fred. "The Quest for Oil." *National Geographic* Vol. 176 (August 1989): 226–263.

Kerr, Richard A. "Extracting Geothermal Energy Can Be Hard." *Science* Vol. 218 (November 12, 1982): 668–669.

Marlay, Robert C. "Trends in Industrial Use of Energy." *Science* Vol. 226 (December 14, 1984): 1277–1282.

Rogers, Michael. "After Oil: What Next?" *Newsweek* (June 30, 1986): 68–69.

INDUSTRIAL POLLUTION

Crawford, Mark. "Hazardous Waste: Where to Put It?" *Science* Vol. 235 (January 9, 1987): 156–157.

Idso, S. B. "Industrial age leading to the greening of the earth." *Nature* Vol. 320 (March 6, 1986): 22.

Kao, Timothy W. and Joseph M. Bishop. "Coastal Ocean Toxic Waste Pollution: Where Are We and Where Do We Go?" *USA Today* Vol. 114 (July 1985): 20–23.

Maranto, Gina. "The Creeping Poison Underground." *Discover* Vol. 6 (February 1985): 75–78.

Monastersky, Richard. "Fate of Arctic ozone remains up in the air." *Science News* Vol. 135 (January 21, 1989): 37.

Raloff, Janet. "U.S. river quality: Not all signs are good." *Science News*, Vol. 131 (April 4, 1987): 214–215.

Sun, Marjorie. "Ground Water Ills: Many Diagnoses, Few Remedies." *Science* Vol. 232 (June 20, 1986): 1490–1493.

Tangley, Laura. "Acid Raid Threatens Marine Life." *BioScience* Vol. 38 (September 1988): 538–539.

DEFORESTATION

Booth, William. "Monitoring the Fate of the Forests from Space." *Science* Vol. 243 (March 17, 1989): 1428–1429.

_____. "Johnny Appleseed and the Greenhouse." *Science* Vol. 242 (October 7, 1988): 19–20.

Cohn, Jeffrey P. "Gauging the biological impacts of the greenhouse effect." *BioScience* Vol. 39 (March 1989): 142–146.

Colinvaux, Paul A. "The Past and Future Amazon." *Scientific American* Vol. 260 (May 1989): 102–108.

Marshall, Eliot. "EPA's Plan for Cooling the Global Greenhouse." *Science* Vol. 243 (March 24, 1989): 1544–1545.

Monastersky, Richard. "Climate influence on forest fires." *Science News* Vol. 134 (July 23, 1988): 55.

Raloff, Janet. "Deforestation: Major threat to ozone?" *Science News* Vol. 130 (August 23, 1986): 119.

Revkin, Andrew. "Cooling off the Greenhouse." *Discover* Vol. 10 (January 1989): 30–33.

Roberts, Leslie. "Is There Life After Climate Change?" *Science* Vol. 242 (November 18, 1988): 1010–1013.

Rock. B. N., et al. "Remote Detection of Forest Damage." *BioScience* Vol. 36 (July/August 1986): 439–444.

Woodwell, G. M., et al. "Global Deforestation: Contribution to Atmospheric Carbon Dioxide." *Science* Vol. 222 (December 9, 1983): 1081–1085.

AGRICULTURE

Batie, Sandra S. and Robert G. Healy. "The Future of American Agriculture." *Scientific American* Vol. 248 (February 1983): 45–53.

Bower, Bruce. "Recasting plaster in Late Stone Age." *Science News* Vol. 134 (October 1, 1988): 213.

Diamond, Jared. "The Worst Mistake in the History of the Human Race." *Discover* Vol. 8 (May 1987): 64–66.

Gibbons, Boyd. "Do We Treat Our Soil Like Dirt?" *National Geographic* Vol. 166 (September 1984): 353–388.

Hinman, C. Wiley. "New Crops for Arid Lands." *Science* Vol. 225 (September 28, 1984): 1445–1448.

Lewin, Roger. "A Revolution of Ideas in Agricultural Origins." *Science* Vol. 240 (May 20, 1988): 984–986.

Maranto, Gina. "A Once and Future Desert." *Discover* Vol. 6 (June 1985): 32–39.

Raloff, Janet. "Salt of the Earth." *Science News* Vol. 126 (November 10, 1984): 298–301.

Stone, Judith. "Bovine Madness." *Discover* Vol. 10 (February 1989): 38–41.

Tangley, Laura. "Preparing for climate change." *BioScience* Vol. 38 (January 1988): 14–18.

THE MELTING ICE CAPS

Beard, Jonathan. "Glaciers on the run." *Science 85* Vol. 6 (February 1985): 84.

Bowen, D. Q. "Antarctic ice surges and theories of glaciation." *Nature* Vol. 283 (February 14, 1980): 619–620.

Hansen, J. V. Gornitz, S. Lebedeff, and E. Moore. "Global Mean Sea Level: Indicator of Climatic Change?" *Science* Vol. 219 (February 25, 1983): 996–997.

Kerr, Richard A. "Climate Since the Ice Began to Melt." *Science* Vol. 226 (October 19, 1984): 326–327.

Matthews, Samuel W. "Ice On The World." *National Geographic*. (January 1987): 79–103.

Mollenhauer, Erik and George Bartunek. "Glacier on the move." *Earth Science* Vol. 41 (Spring 1988): 21–23.

Monastersky, Richard. "Shrinking ice may mean warmer Earth." *Science News* Vol. 134 (October 8, 1988): 230.

Peltier, W. R. "Global Sea Level and Earth Rotation." *Science* Vol. 240 (May 13, 1988): 895–900.

Radok, Uwe. "The Antarctic Ice." *Scientific American* Vol. 253 (August 1985): 98–106.

Zwally, H. Jay, C. L. Parkinson, and J. C. Comiso. "Variability of Antarctic Sea Ice and Changes in Carbon Dioxide." *Science* Vol. 220 (June 3, 1983): 1005–1012.

GLOBAL WARMING

Jager, Jil. "Anticipating Climate Change." *Environment* Vol. 30 (September 1988): 13–15 & 30.

Lovejoy, Thomas E. "Will unexpectedly the top blow off?" *BioScience* Vol. 38 (November 1988): 722–726.

Monastersky, Richard. "Looking for Mr. Greenhouse." *Science News* Vol. 135 (April 8, 1989): 216–221.

_____. "Global Change: The Scientific Challenge." *Science News* Vol. 135 (April 15, 1989): 232–235.

Ramanathan, V. "The Greenhouse Theory of Climate Change: A Test by an Inadvertant Global Experiment." *Science* Vol. 240 (April 15, 1988): 293–299.

Revelle, Roger. "Carbon Dioxide and World Climate." *Scientific American* Vol. 247 (August 1982): 35–43.

Revkin, Andrew C. "Endless Summer: Living With the Greenhouse Effect." *Discover* Vol. 9 (October 1988): 50–61.

Schneider, Stephen H. "The Greenhouse Effect: Science and Policy." *Science* Vol. 243 (February 10, 1989): 771–779.

_____. "Climate Modeling." *Scientific American* Vol. 256 (May 1987): 72–80.

Wigley, T. M. L., P. D. Jones, and P. M. Kelly. "Scenario for a warm, high CO_2 world." *Nature* Vol. 283 (January 3, 1980): 17–21.

Woodwell, George M. "Contribution to Atmospheric Carbon Dioxide." *Science* Vol. 222 (December 9, 1983): 1081–1085.

_____. "The Carbon Dioxide Question." *Scientific American* Vol. 238 (January 1978): 34–43.

Usher, Peter. "World Conference on the Changing Atmosphere: Implications for Global Security." *Environment* Vol. 31 (January/February 1989): 25–27.

Index

About the Author

Jon Erickson has written several books on earth science for TAB. He holds an advanced degree in natural science and has worked as a geologist for major oil and mining companies and as an engineer for an aerospace company. He presently makes his home in western Colorado, where he works as an independent geologist and writer.

TAB books by the author:

- *Volcanoes and Earthquakes* (No. 2842)
- *Violent Storms* (No. 2942)
- *The Mysterious Oceans* (No. 3402)
- *The Living Earth: The Coevolution of the Planet and Life* (No. 3142)
- *Exploring Earth from Space* (No. 3242)